Deutsche
Forschungsgemeinschaft

**Status and Prospects
of Astronomy in Germany
2003–2016**

Memorandum

Deutsche Forschungsgemeinschaft

Status and Prospects of Astronomy in Germany 2003–2016

Memorandum

Editorial committee:
Andreas Burkert, Reinhard Genzel, Günther Hasinger, Gregor Morfill (Chair), Peter Schneider, Detlev Koester (Chair of the Council of German Observatories)

WILEY-VCH Verlag GmbH & Co. KGaA

Deutsche Forschungsgemeinschaft (DFG)
German Research Foundation
Head Office: Kennedyallee 40, 53175 Bonn
Postal address: 53170 Bonn
Tel. +49 228 885-1
Fax +49 228 885-2777
postmaster@dfg.de
www.dfg.de

> The work presented here was compiled with great care. Nevertheless, the authors, editors and publisher cannot accept responsibility for the correctness of information, notes and advice, or for any printing errors.

Bibliographic information published by Die Deutsche Bibliothek
Die Deutsche Bibliothek lists this publication in the Deutsche Nationalbibliografie; detailed bibliographic data is available on the internet at <http://dnb.ddb.de>.
ISBN: 978-3-527-31910-7

© 2003 WILEY-VCH Verlag GmbH & Co. KGaA, Weinheim

© englische Übersetzung 2008 WILEY-VCH Verlag GmbH & Co. KGaA, Weinheim

Printed on acid-free paper.

All rights reserved (including those of translation into other languages).
No part of this book may be reproduced in any form - by photoprinting, microfilm, or any other means - nor transmitted or translated into a machine language without written permission from the publishers. Registered names, trademarks, etc. used in this book, even when not specifically marked as such, are not to be considered unprotected by law.

Cover design and typography: Dieter Hüsken.
Typesetting: Hagedorn Kommunikation, Viernheim.
Printing: betz-druck GmbH, Darmstadt.
Binding: Litges & Dopf GmbH, Heppenheim.
Printed in the Federal Republic of Germany.

Contents

Editorial committee VII

Translation preface IX

Preface to the original memorandum XIII

Summary XV

1	**Astronomy yesterday, today and tomorrow**	1
1.1	The history of the universe	1
1.2	The life cycle of stars and the matter cycle	7
1.3	New windows into space	17
2	**The scientific issues**	23
2.1	The universe – its origin, evolution and large-scale structure	23
2.1.1	The big bang and cosmic background radiation	25
2.1.2	The cosmic distance scale	28
2.1.3	Cosmological evolution	32
2.1.4	The large-scale structure of the universe	37
2.1.5	Galaxy clusters as cosmic laboratories	41
2.1.6	Ultra-high-energy gamma-ray astronomy	45
2.1.7	Astroparticle physics	47
2.1.8	New properties of neutrinos	50
2.1.9	Gravitational wave astronomy	52
2.2	Galaxies and massive black holes	53
2.2.1	Origin and evolution of galaxies	56
2.2.2	Structure of galaxies	65
2.2.3	Massive black holes	70
2.2.4	Active galactic nuclei	77
2.3	The matter cycle and stellar evolution	82
2.3.1	The interstellar medium	84
2.3.2	Cosmic rays	90
2.3.3	The nearest star: the sun	92

2.3.4	The stars.	94
2.3.5	Stars as chemical factories and the engines of the matter cycle	99
2.3.6	The final stages of stellar evolution	102
2.4	Stellar and planetary formation: protostars, circumstellar disks and extrasolar planetary systems	109
2.4.1	Star formation as a fundamental cosmological process	109
2.4.2	Formation of low-mass stars: from prestellar core to dust disk	112
2.4.3	Massive stars, stellar clusters and initial mass distribution	121
2.4.4	Extrasolar planets	124
3	**The next fifteen years: observatories and instruments**	129
3.1	Access to telescopes and involvement in large international projects	129
3.2	Safeguarding competitiveness	137
3.3	Strengthening national innovation and initiative	139
3.4	Other planned space and balloon missions	145
3.5	New initiatives: astroparticles and gravitational wave research	147
3.6	The next decade's projects	150
3.7	The future role of existing establishments	156
3.7.1	Institute for Radio Astronomy in the Millimetre Range (IRAM)	156
3.7.2	The Effelsberg radio telescope	158
3.7.3	Calar Alto observatory	159
3.7.4	Solar telescopes	161
4	**Astronomical research structures**	163
4.1	Historical development	163
4.2	The current situation	166
4.3	Research institutes	168
4.4	Research funding instruments	173
4.5	Training	176
4.6	Astronomy and the public	177
5	**Recommendations**	180
5.1	Some fundamental aspects	180
5.2	Instruments and projects	182
5.3	Large projects	185
5.4	Medium-sized projects	187
5.5	Small projects	187
5.6	Organisational instruments	192
5.7	Interaction and cooperation	195
5.8	Securing and reinforcing funding instruments	198
6	**Annex**	207
	Member institutes of the German Council of Observatories	207
Glossary/Acronyms		218
Telescopes, instruments, experiments		221

Editorial committee

Prof. Dr. Andreas Burkert, Munich
Prof. Dr. Reinhard Genzel, Garching
Prof. Dr. Günther Hasinger, Garching
Prof. Dr. Gregor Morfill (Chair), Garching
Prof. Dr. Peter Schneider, Bonn

Prof. Dr. Detlev Koester, Kiel
(Chair of the Council of German Observatories)

We would like to thank all members of the Council of German Observatories and the Deutsche Forschungsgemeinschaft (DFG) for their energetic support.

From the Head Office of the DFG:
Dr. Stefan Krückeberg
(Programme Director Astronomy and Astrophysics)

Translation Preface

About five years ago, the Deutsche Forschungsgemeinschaft (German Research Foundation, DFG) published the original German version of the memorandum "Status and Prospects of Astronomy in Germany 2003–2016". This report, the third in a series of strategic planning documents for German astronomy, identified priorities for the next decade(s) and the tools and instruments necessary to achieve these objectives. Due to increasing global cooperation and the ensuing need to coordinate planning in an international, and in particular European context, the DFG now presents an English translation of its memorandum to disseminate its findings and facilitate collaborative efforts in this area.

This translation also provides a unique opportunity to reflect on the progress achieved in the different areas since publication of the memorandum. Fortunately, the basic conditions for public research funding in Germany have improved significantly over the last years, as the major German research organisations have profited from a steady increase of their budgets. In addition, the Excellence Initiative launched in 2005 makes a significant contribution to strengthening science and research in Germany in the long term, improving its international competitiveness and raising the profile of the top performers in academia and research. Also the German national space programme, managed by the German Aerospace Center, DLR, enjoyed a significant budgetary increase allowing the start of new projects in basic space research.

The focussing of scientific priorities in the memorandum helped the German astrophysical communities to develop strategic coordinated efforts, which can be seen in the newly established collaborative DFG projects: a Priority Programme entitled "Witnesses of Cosmic History – Formation and Evolution of Galaxies, Black Holes, and Their Environment" is a nationwide effort; the Transregional Collaborative Research Centre "The Dark Universe" combines the expertise of scientists from Heidelberg, Bonn and Munich; an interdisciplinary team of astrophysicists and mineralogists investigates "The Formation of Planets – the Critical First Growth Phase" in a Research Unit; "Gravitational Wave Astronomy" is the subject of a Transregional Collaborative Research Centre involving experts from Jena, Tübingen, Hanover, Potsdam and Munich; the new Research Training Groups on "Theoretical Astrophysics and Particle Physics" (Würzburg) and on "Extrasolar Planets and their Host Stars" (Hamburg, Göttingen, Katlenburg) complement the picture.

Astrophysics is also a vital ingredient in several successful projects in the framework of the Excellence Initiative. These include the Excellence Cluster "Origin and Structure of the Universe" (Munich), the Excellence Cluster "QUEST: Centre for Quantum Engineering and Space-Time Research" (Hanover), the "Heidelberg Graduate School on Fundamental Physics" and the "Bonn-Cologne Graduate School of Physics and Astronomy".

The Cooperative Research System (Verbundforschung) for astronomy and astrophysics is a federal project funding line that was established as a result of the 1987 DFG astronomy memorandum. It enables German universities in particular to participate in the instrument development and scientific utilisation of large projects co-funded by the German government. In the course of preparing the 2003 memorandum it was possible to secure this funding line and extend it with a new channel for astroparticle physics.

Despite the new opportunities opening up in this field, one of the main concerns as stated in the memorandum, namely the structural weakness of astrophysics at German universities, has still not seen a significant improvement and will probably take substantial time to resolve. In part, the Excellence Initiative is helping here, but a much broader approach at a larger number of universities is required.

The priorities set out in the memorandum have already influenced further strategic planning processes in an international context, for instance the European Southern Observatory (ESO) long-range plan or the European Space Agency (ESA) Cosmic Vision exercise. Currently, the European Research Area Networks (ERA-Net) Astronet and ASPERA, organised by European funding agencies and funded by the European Commission, are putting together a European science vision and infrastructure roadmap for astrophysics and astroparticle physics.

The highest priority projects outlined in the report continue to be topical today. The following developments have occurred since publication of the German report:
- The Atacama Large Millimeter Array (ALMA) is in construction in a global collaboration between Europe, the United States and Japan. Its precursor, the German APEX telescope, is producing interesting scientific findings.
- The Herschel ESA Space Telescope, with major German instrument contributions, will be launched in 2008.
- The Cherenkov telescopes H.E.S.S. and MAGIC are both producing excellent scientific results and are being upgraded with additional telescopes.
- The ICECUBE neutrino detector in the Antarctic ice is under construction.
- LISA, the Laser Interferometer Space Antenna, a joint ESA/NASA observatory to detect gravitational waves, has been selected as part of the ESA Cosmic Vision 2015–2025 programme, together with its technological precursor, LISA Pathfinder, which is planned for launch in 2010.
- SOFIA, the airborne far-infrared telescope project developed jointly by NASA and the DLR, has been continued; the airplane has performed several test flights and scientific operations are scheduled to start in 2009.
- Solar Orbiter, an ESA solar observatory, is a confirmed part of the ESA Cosmic Vision programme with a launch expected in 2015.

- The ESO Very Large Telescope (VLT), with its interferometer VLTI, has now become the internationally leading astronomical installation, and second-generation instruments are being prepared with significant German contributions.
- The Large Binocular Telescope (LBT) in Arizona has been commissioned and is delivering exciting science with both telescopes. A full suite of instruments is being prepared and will be available in the next few years.
- The X-ray Evolving Universe Spectroscopy mission XEUS, a large aperture X-ray mirror in formation flight with a separate detector spacecraft, has recently been selected for an assessment study in the ESA Cosmic Vision programme. The intermediate X-ray astronomy project eROSITA has been approved by the DLR and is scheduled for launch in 2011.

In the course of new international strategic planning processes, in particular the European ESFRI roadmap for large research infrastructures and the Astronet and ASPERA ERA-Nets, new large-scale projects with significant German interests are being prepared or discussed, some of which have already been mentioned in this memorandum. One notable example is the E-ELT, a very large European optical/near-infrared 42m-telescope (previously called OWL), which recently advanced to a phase B study. In the radio range, the Square Kilometer Array (SKA), a very large centimetre-radio array has been proposed and its precursor, LOFAR, is currently being built between the Netherlands, Germany and the United Kingdom. Simbol-X, a formation-flying hard X-ray telescope with high resolution focussing optics, is a precursor of XEUS, studied jointly by France and Italy with key German contributions. On the astroparticle side two new projects are being discussed: the Cherenkov Telescope Array (CTA), a natural successor of H.E.S.S. and MAGIC, and KM3NET, a water Cherenkov neutrino telescope in the Mediterranean Sea. In the current European funding structure it will not be possible to support all these next generation projects in parallel, so that either a phased approach and/or global collaborations have to be implemented.

Considering the impact the memorandum has made in recent years, it is clear that the English translation will further help to keep the global research community abreast of current priorities and developments, serving as a tool to help foster collaboration in this exciting field.

Bonn, April 2008

Prof. Dr.-Ing. Matthias Kleiner
President of the DFG

Preface to the original memorandum

Modern astronomical research occupies itself with questions so fascinating nobody can escape them: how did the universe form, how is matter distributed within it, what will be its final destiny? How do stars, galaxies and black holes form? Under what conditions can planetary systems form around stars? The advances achieved in the past two decades in answering these questions are enormous and have spurred intense public interest. At the same time, technological innovations have opened new windows into space, promising unimaginable new discoveries for the future. Current astronomy research is very heavily dominated by physics, making astronomy and astrophysics synonymous.

The "Astronomy Memorandum" presented here identifies the most important astrophysical research priority programmes for the next 10 to 15 years in the shape of an inventory and presents the requirements for allowing German astrophysicists to play an internationally leading role in the future, as they had in the past. It was compiled by the Council of German Observatories and carries on in the tradition of the memoranda of 1962 and 1987. First, the great scientific successes of the past are sketched out and then the core astrophysical questions of the next two decades described. The memorandum is therefore aimed at researchers in universities and in non-university establishments, whose mutual dedication is necessary in order to contribute to answering these questions. The priorities for the use of existing, and the building of new, generally international, observation facilities are defined on this basis. Further recommendations involve the structure of astronomy research in Germany. The memorandum is therefore also aimed at political offices on a federal and state level.

Coordinated interaction of all national and international partners plays a vital role in the optimum funding of astronomy. Among other things, the 1987 DFG astronomy memorandum led the way for the establishment of cooperative research by the BMBF. In the fields of ground-based astronomy, satellite-supported astronomy and astroparticle physics, the project-specific use of large international instruments and participation in their instrumentation are supported.

This important funding instrument has borne rich fruits in the last decade and represents a direct success of the previous memoranda. It is to be hoped that the measures presented in this memorandum are implemented as far as possible and with a long-term planning horizon, allowing German scientists to continue making successful contributions to the development of astronomy.

I would like to thank the authors and all those who have contributed to this impressive memorandum by their discussions and suggestions.

Bonn, November 2003

Prof. Dr. Ernst-Ludwig Winnacker
President of the DFG

Summary

Astrophysics is currently one of the most exiting fields in physics and will probably remain so for the foreseeable future. New telescopes operating in all wavelength ranges, extremely sensitive detectors and the opening of completely new observation windows into space promise enormous potential, associated with new challenges for theoretical astrophysics. Researchers in Germany play a major role, often even a leading role, in this heavily internationally linked field. Now – 15 years after the last German "Astronomy Memorandum"[1] and at the beginning of a new century – is a good time to portray the advances made in this rapidly growing field, to assess the most important directions and developments of the coming 10 to 15 years and to develop conclusions in terms of German astrophysics. That is the purpose and aim of this memorandum, compiled by the German astronomy community (under the collective umbrella of the Council of German Observatories) in coordination with the Deutsche Forschungsgemeinschaft.

The questions of the origin and the evolution of the universe, the objects it contains and the physical laws determining their behaviour are at the centre of astrophysics. Over the past few years there have been decisive breakthroughs and paradigm shifts on a number of fronts, so that the current period can be rightly seen as a "Golden Age" of astrophysics. One decisive breakthrough, for example, was made in understanding the expansion of the universe. The big bang inflation theory allowed concrete predictions, which have been recently confirmed with a great degree of precision, for example about minute quasi-periodic spatial oscillations in the microwave background radiation. From this, from observing very remote supernova explosions, and from X-ray surveys of galaxy clusters, it was possible to derive the geometry of space and the mean density of the cosmos. This has shown that the matter in the cosmos is dominated by a previously unknown particle type, which is known as "dark matter".

The discovery that all known matter is not sufficient to close the universe was somewhat surprising. It means that the cosmos will probably continue to expand into eternity. Even more unexpected was the discovery that the expansion of the universe

---

[1] Astrophysics and astronomy are used synonymously here. Cosmology forms a subset of astrophysics, involved with the origin and evolution of the universe.

is still accelerating. This indicates the presence of a previously completely unknown "dark energy", which dominates the universe, today and in the future. Cosmology on the largest scale is intimately connected to the physics of the smallest scale. It is the discoveries made in astrophysics during the last decades that represent the driving force behind extending the boundaries of the particle physics standard model (for example the discovery of neutrino oscillations). The greatest challenge today is the unification of the theory of relativity with quantum theory. New discoveries in cosmology, on the other hand, represent stringent tests for modern particle theories (for example the superstring and Brane-World theories) and may point the way to a new understanding of the vacuum energy.

A paradigm shift has also taken place with respect to black holes, which are often regarded as highly exotic, possibly purely theoretical constructions. In the meantime, objects that very probably represent black holes and for which all alternative explanations are considerably more exotic, have been discovered and investigated closely. In our Milky Way we know of black holes of only a few solar masses, so called stellar black holes. A massive black hole of several million solar masses has been identified at the centre of our Milky Way. In addition, massive black holes exist in the centres of almost all large, nearby galaxies. Statistical investigations show that these black holes must have originated in the very early universe, probably together with the first stars and galaxies. They are therefore fundamentally connected to our existence. A further breakthrough involved the discovery of extrasolar planets. Today, we know of more than 100 planetary systems outside of our own solar system and this number increases by the week. Among others, a planetary system that may be similar to our solar system was discovered recently. In about a decade, we hope to use telescopes that are sensitive enough to discover earthlike worlds in different solar systems, perhaps even discovering signs of life on other planets.

The most important research topics of the future will be concerned with the origin and evolution of the universe as a whole, of galaxies and massive black holes, and of stars and planetary systems. Concrete tasks here include identifying the exact geometry of the universe; the nature of dark matter and dark energy; the discovery of gravitational waves, the first galaxies, the first black holes and the first generation of stars; the origin of massive stars, stellar explosions and the merging of compact objects; as well as unravelling the nature of gamma ray bursts, the origin of planetary systems and, last but not least, the search for biological activity on extrasolar planets.

Two topics will serve as examples for discussion here. They will demonstrate how the development of new observation options goes hand in hand with attempts to answer fundamental astrophysical questions.

In order to understand the formation and evolution of galaxies in the early universe, we require extremely sensitive observations in the relatively long-wave ranges of the electromagnetic spectrum – and thus of the cold universe. The light of the stars in the earliest galaxies is shifted towards the near- to mid-infrared due to their relative motion away from us. Beginning in 2005 the German-American airborne observatory SOFIA *(Stratospheric Observatory for Infrared Astronomy)*, in 2007 the European Cornerstone mission Herschel and in 2009/2010 NASA's and the European Space Agency's *James Webb Space Telescope* (JWST) will carry out increasingly sensitive observations[2]. Parallel to this, ground-based adaptive optics and interferometry will make extremely high resolution and sensitive images of remote galaxies in the near infrared, first using the ESO's *Very Large Telescope Interferometer* (VLTI) and the international *Large Binocular Telescope* (LBT), in the next decade perhaps using large telescopes of 50 m to 100 m diameter. But we have also known for a number of years that a large part of the light from the first stars is blocked by dense dust clouds in the early galaxies and is then re-emitted in the far infrared and submillimetre range. Herschel is also highly sensitive in the far infrared range, but it is the millimetre and submillimetre interferometer ALMA *(Atacama Large Millimeter Array)*, developed in global cooperation, which will spatially resolve dust and gas masses in even the most remote galaxies and investigate their structure and dynamics. Its total of 64 antennas will be erected by 2010 on the Chajnantor Plateau in Chile. Thanks to its unique combination of angular resolution, spectral resolution and sensitivity, ALMA will be capable of studying in detail stellar nurseries in our Milky Way with their protostars and protoplanetary disks, as well as – in cooperation with X-ray observations – hidden black holes.

The second topic concerns itself with the hot, energetic universe and opening new observation windows into space. The 2002 Nobel Prize for Physics was awarded for "pioneering work in astrophysics", with half going to Raymond Davis Jr. and Masatoshi Koshiba for the discovery of cosmic neutrinos, and half to Riccardo Giacconi for his contributions to the discovery of cosmic X-ray sources. The discoveries made by Davis and Koshiba have gradually opened up the new field of neutrino astronomy, with great consequences for particle physics, astrophysics and cosmology. We hope to be able to carry out routine astronomy using new, large neutrino telescopes within the next 15 years, for example with *ICECUBE* in the Antarctic. The X-ray astronomy initiated by Giacconi showed for the first time that certain objects in our universe convert enormous amounts of energy in a very short time and thus opened a window onto the most extreme states of matter: dying stars, neutron stars and black holes. In the 40 years or more of its existence, X-ray astronomy has become an integral and indispensable part of astrophysics, primarily with German and European assistance (for example by way of the *ROSAT* X-ray satellite). Among other things, the challenges of the future lie in precisely surveying the strong gravitational field close to black holes and in the discovering of the earliest black holes in the universe.

It is for this purpose that, together with Japan, ESA is planning the *X-Ray Evolving Universe Spectroscopy Mission* (XEUS), a telescope consisting of two satellites separated by 50 metres. Their mirrors will be assembled from several individual components by astronauts on the International Space Station, some time after 2012[3]. Gravitational wave astronomy also involves opening a new window to extreme forces and states of matter. In 1993 the American astrophysicists Joseph Taylor and Russell Hulse were awarded the Nobel Prize for Physics for their indirect proof of the existence of gravitational waves on two neutron stars in mutual orbit. Direct proof

---

[2] The start times of operations have been substantially delayed: SOFIA to 2009, Herschel to 2008 and JWST to 2013.

[3] In the meantime it is planned to launch a 4.5 m diameter mirror and a separate detector spacecraft directly into a deep space orbit.

Summary

of gravitational waves using large laser interferometer detectors, a number of which are already operating, is hoped for within the next fifteen years. Particularly important here is the *Laser Interferometer Space Antenna* (LISA), planned jointly by ESA and NASA, a combination of three satellites spaced 5 million kilometres apart. From around 2011[4] they will routinely observe gravitational waves produced by merging, massive black holes throughout the universe.

The principal astrophysics funding instruments in Germany have proved their usefulness and should be retained or extended. The division of tasks between the various funding organisations has established itself over a long period, among other things as the result of past memoranda. The largest astrophysics projects (for example ALMA or LISA) can only be realised within a framework of international, or even global, cooperation. Germany is integrated in the European space- and ground-based astronomy (ESA and ESO) organisations as one of the most important partners. Large national or transnational observatories (for example LBT, ROSAT), but also instruments for these large international projects, can only be carried by large national institutes with the appropriate fundamental equipment and infrastructure, available for long-term planning: Max Planck Institutes, Leibniz Institutes, state institutes and a few university institutes. Project funding by the DLR within the national space programme, BMBF cooperative research in the ground-based field and funding by coordinated DFG processes such as Collaborative Research Centres is indispensable. Use of large observatories, both on the ground and in space, is accessible to all researchers within the framework of scientific competition. The necessary funding is provided by way of basic institutional funding, the proven funding structures of the DFG, and DLR cooperative research for the space observatories. The most important structural recommendations of the memorandum are summarised below.

Further involvement in the large European research organisations ESO (for ground-based astronomy) and ESA (for space-based astronomy), and national scientific use of, or instrumentation development for, the observatories supported by these organisations enjoys highest priority.

Essential to this is sufficient and reliable basic funding by the Max Planck Institutes, Leibniz Institutes and state institutes, as well as increased utilisation of the proven funding options provided by the DFG, which in turn requires sufficient overall funding of the DFG.

A noticeable improvement in the infrastructure of the university astronomy institutes is essential, especially in terms of personnel, the establishment of new professorships and expansion of the subjects offered for study, leading to nationwide astronomy education, including for prospective physics teachers. Inter-state cooperation in particular requires further underpinning.

In the mid-term, funding for the national extraterrestrial programme, which has been declining for more than ten years, must be stabilised and stocked up again. It is indispensable for the competitive utilisation of large investments with regard to ESA and to retaining national innovative capacity and technological expertise. The considerable development and project running times make long-term planning necessary.

---

[4] The earliest launch time for LISA is now 2018.

Summary

The very successful BMBF/DLR astronomy/astrophysics cooperative research element should be retained and expanded. Cooperative research was approved in an agreement between DFG, BMBF and DLR in the course of the last memorandum and was extended by the astroparticle physics field during the preparation for this memorandum. It is absolutely necessary if large investments in ground- and space-based astronomy are to pay off scientifically on a wide front, in particular at the universities.

The sun is "our" star and therefore naturally also forms a component of this "Astronomy" memorandum. However, solar system research has been generally ignored here. Other topics, some with many similarities to astrophysics, were also neglected in order to concentrate on the core issues. These include stellar particle winds, planetary magnetospheres, interplanetary dust, planetary rings, meteorites, asteroids, comets, planets and moons, and cosmochemistry data. In the first chapter, the memorandum offers a general introduction to current developments in astrophysics. In the second chapter, four research priorities are described, representing the great challenges and probably the most important topics of the future. The third chapter introduces the observatories and instruments of the next 15 years. In the fourth chapter, the structure of astronomy research in Germany is discussed, and in the fifth chapter the most important observatories and instruments are prioritised and the structural recommendations discussed.

1 Astronomy yesterday, today and tomorrow

1.1 The history of the universe

The universe – space, time and all matter – suddenly came into being around 14 billion years ago from an extremely hot and dense state. Ever since this "big bang", space and its embedded matter have been expanding. The reason we are able to reconstruct the history of the universe is related to the finite travel time of light: each look into the depths of space is also a look into the deep past.

The echo of the big bang and determination of cosmological parameters

A directly observable trace of the earliest phase of the universe is the cosmic microwave background radiation. This remnant also represents the primary pillar of the big bang theory. About 380,000 years after the big bang, matter in its gaseous form had cooled enough to allow atomic nuclei and electrons to combine and form atoms, consisting almost entirely of hydrogen and helium. Matter became transparent to light during this era and released radiation began to fill the universe. It can still be detected today as a radiation field that uniformly fills the sky. Due to the expansion of the cosmos this radiation field has cooled to 2.7 K, which is why we refer to this as the 3K radiation.

The discovery of this homogeneous microwave background in 1965 by Penzias and Wilson was rewarded with the Nobel Prize for Physics. It was not possible to identify any underlying structures, known as anisotropies, until 1992 with the American led COsmic Background Explorer (COBE) telescope. These features are interpreted as overdensities in the primordial gas. Large-scale structures of the universe, such as galaxy clusters, later arose from them (Figure 1.1). Investigations of the structure of the microwave background, together with numerous other studies, provide consistent values for the cosmological parameters, which describe the geometry

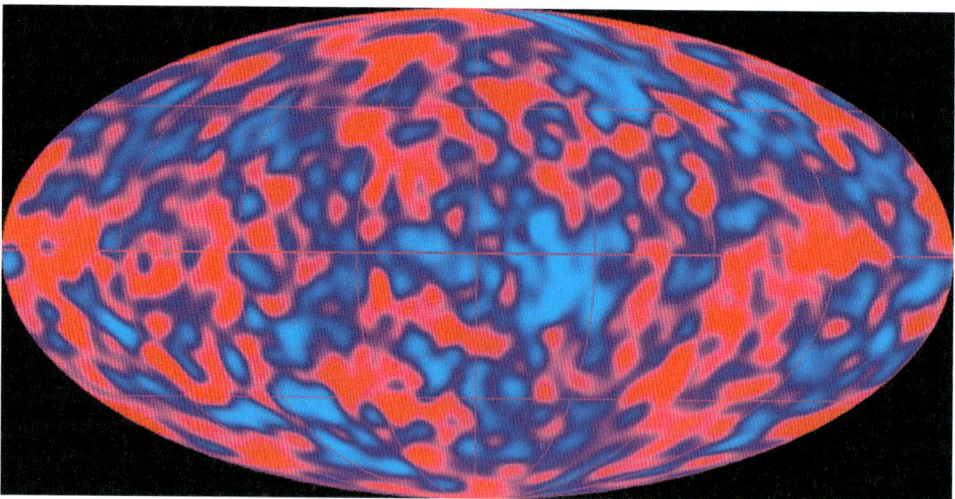

Fig. 1.1: Structures in the cosmic background radiation viewed at an angular resolution of seven degrees. The speckled pattern shows minor fluctuations in temperature, originating from the initial overdensities in the primordial gas. (NASA and the COBE Science Working Group/ESA)

and extent of the universe. They include the Hubble constant, describing the expansion and age of the observable universe, and the density of the various components of matter, thus determining the curvature and expansion history of space.

COBE has an angular resolution of only seven degrees, corresponding to the diameter of 14 full moons, and could therefore only give a rough picture of these "condensation seeds." However, theoretical considerations require that it is precisely these smaller structures which contain further key information about the conditions prevalent during the big bang, and about the creation and further development of the universe. For this reason, one of the central tasks required is the study of the 3K background radiation with far smaller angular resolution and greater sensitivity than was possible with COBE. In the interim, additional measurements were made using high altitude research balloons to proceed further. A breakthrough was achieved by NASA's *Wilkinson Microwave Anisotropy Probe* (WMAP) mission, which confirmed with great precision the projections of the standard big bang model thus determining the cosmological parameters with accuracy. However, comprehensive information on the polarisation and foreground microwave radiation emission will not be recorded until the Planck satellite, developed by the European Space Agency (ESA) to have three times greater resolution, becomes available.

1.1 The history of the universe

In the distant future, it may yet be possible to identify a cosmic neutrino or even a gravitation wave background, allowing us to view into even earlier epochs of our universe.

Dark matter and dark energy

Probably the most mysterious component of the standard cosmological model is dark matter. It is inferred to exist in many regions of space due to its gravitational effects, but cannot be detected by the emission or absorption of radiation. According to the most recent research results, it represents around one third of the total energy density of the universe. In comparison, the chemical elements of "normal" matter, or what is known as baryonic matter, of which all visible objects are made, contributes only a few percent. Most of the cosmic energy density appears to be composed of this recently discovered and generally mysterious "dark energy". These findings point to a heightened acceleration of cosmic expansion.

From a theoretical point of view, it is assumed that dark matter first formed the overdensities in the early universe, then their gravitational fields collected normal matter that condensed to form galaxies. It therefore played a vital role in the development of structure in the universe. Theoreticians try to simulate this process using powerful computers (Figure 1.2) though the composition of dark matter remains generally unclear. Only a small fraction of it can be made out of faint sky objects. From theoretical considerations, it is probable that we are dealing with an unknown class of elementary particle. This paves the way for a fascinating alliance between cosmology and particle physics.

One promising means of detecting dark matter and determining its mass distribution is provided by the gravity lensing effect. Here, matter, for example in a galaxy cluster, curves the surrounding space so heavily that light is distorted as if in a lens. The image of a galaxy hidden behind such a cluster is then bowed to form a circular arc (Figure 1.3). The mass of the dark matter and its spatial distribution can be determined from these images with the aid of models.

Recent research results come to the conclusion that the remaining cosmic energy density is a form of vacuum energy (corresponding to the cosmological constant of Einstein's Theory of Relativity). This has played a decisive role in the development of the universe. The presence of this mysterious dark energy leads to an accelerated expansion of the universe. While models exist that may clarify the nature of dark matter from the perspective of elementary particle physics, the existence of dark energy comes as a complete surprise. Understanding this type of matter will ultimately lead to

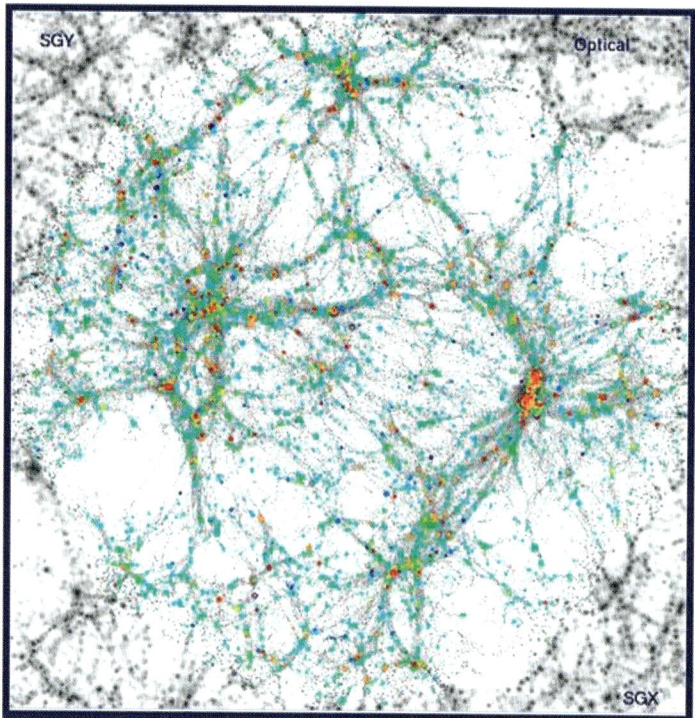

Fig. 1.2: The most powerful computers in the world are required to model the creation of galaxies and galaxy clusters from the primordial gas. The coloured points represent various galaxy types, shown by the colour of their star populations, while the grey background reveals the dark matter. Elliptical galaxies with red, older star components are generally created in galaxy clusters, while in the filaments, spiral galaxies with luminous, young blue stars are primarily formed. (MPA)

Fig. 1.3: The entire matter in a galaxy cluster acts as a gravitational lens and distorts the image of far-distant galaxies to a circular arc. (ESO)

new, fundamental insights into the microphysics and origin of matter.

The first galaxies formed from overdensities (i.e., condensation seeds) in the primordial gas. When exactly this era began and precisely which processes played a role are crucial questions posed by modern research. It is highly probable that the quasars came into being at the same time as the galaxies. They form in extremely compact and luminous central regions of galaxies. As far as we know today, a black hole of several billion solar masses is located at the centre of every quasar. Black holes attract gas and stars in their neighbourhood and devour the matter they contain. During this process, also known as accretion, up to ten thousand times more energy is released in a region the size of our solar system than is radiated by all the stars in the Milky Way thus allowing quasars to be identified at very great distances. The record holders for the most distant objects are quasars that we see as they were when the universe was less than a billion years old.

The first galaxies and quasars

Quasars have been recently found to play a more important role during the early phase of the universe than was previously suspected. For example, using the German-developed ROSAT X-ray space telescope, astrophysicists discovered that the X-ray background radiation, first identified around 40 years ago, originates from innumerable quasars at early stages of the universe. In 2002, Riccardo Giacconi was awarded the Nobel Prize for Physics due to his pioneering work on the construction of the first X-ray telescopes. These opened the window to the field of X-ray astronomy and to the discovery of background X-ray radiation and its sources.

The existence of massive black holes in the centres of very many, perhaps even all, galaxies is now regarded as fact. The 1990s saw the introduction of an abundance of new data. One important contribution comprises the measurements carried out by German astronomers of the movements of stars within just a few light days of the centre of the Milky Way.

For the first time it was possible to observe with high precision a complete orbit of a star circling the central black hole. The high velocity of this star indicates a dark, central concentration of mass. It is highly probable that this is a black hole of approximately three million solar masses. It has only recently been discovered that the mass of central black holes correlates very well to the velocities of the stars in the mother galaxies. This is indicative of a common creation process for both the black hole and its surrounding galaxy.

The decisive breakthrough, in the search for the most distant and therefore youngest galaxies and quasars, has only been made during the last five years by combining the capabilities of the 10 m Keck telescope and the Hubble Space Telescope. German research groups are now in a position to contribute to this field of research, thanks to the *Very Large Telescope* (VLT) at the European Southern Observatory (ESO).

Formation of galaxies and massive black holes

The most recent data indicates that the early history of young galaxies must have been very turbulent. Observations made using the European Infrared Space Observatory (ISO) verify that the creation of new stars must have been quite explosive in many young galaxies.

The galaxies were also closer together than they are today. Galaxies often collided and merged with one another. Observations of nearby galaxies have shown that these collisions initiated vigorous phases of star production. Massive black holes, observed in many galactic centres, may have also formed during this phase and subsequently been fed by the gases they attract.

Great expectations are linked to observations in the infrared to millimetre range which may facilitate our understanding of the formation and early development of galaxies. For example, the German-American airborne observatory SOFIA *(Stratospheric Observatory for Infrared Astronomy)* is expected to start operations in 2005. The European space telescope Herschel will also create a new benchmark for observations in the submillimetre range. The launch is planned for 2007. From 2010 onwards, even the most remote galaxies will be spatially resolved, and their structures and dynamics studied using ALMA, the interna-tional millimetre and submillimetre interferometer in Chile. Finally, the joint NASA and ESA *James Webb Space Telescope* (JWST) (also from 2009/2010) will be in a position to carry out highly sensitive observations of the first galaxies in the near and mid-infrared.

Active galaxies

Black holes not only played a role in the creation of the galaxies. They are also the root cause of the activity of galactic nuclei. During the past two decades it has been possible to describe the initially very large number of different types of active galaxies using one and the same physical model. This tells us that there is a massive black hole drawing in matter from its surroundings at the centre of every galaxy. Matter first orbits the hole and forms a disk that increases in temperature towards the centre; finally, it falls into the black hole. At this stage the hot gases radiate primarily in the X-ray and UV ranges, as well as emitting visible light.

In many cases, two gas jets are ejected in opposite directions into space at almost light speed and perpendicular to the plane of the disk. The front edges culminate in giant plasma bubbles. They emit bundled radio, gamma, and X-ray radiation. German researchers have contributed substantially to the investigation of these jets and accretionary disks both in theory and in practical observations. For example, it was the Effelsberg radio telescope, a central element of the intercontinental radio interferometric network, that made possible a spatial resolution of these jets up to a few light years distance from black holes. During the last few years gamma ray astronomers have discovered that these jets are also the sources of high-energy photons, whose origin is still not completely clear. In the future, these processes will be studied using special telescopes, such as H.E.S.S. and MAGIC, to identify high-energy photons in the gamma wave range.

1.2 The life cycle of stars and the matter cycle

Our solar system forms part of the Milky Way, a spiral galaxy with a diameter of around 100,000 light years. Studies of our galaxy, its creation and development, represent one of the primary responsibilities of astrophysics. The essential areas of this field include calibration of the absolute distance scale and identification of the fundamental physical stellar parameters.

The Milky Way

Hipparcos, the European astrometric satellite, which finally started in 1989 with the active participation of German astronomers, provided essential contributions in this field. It pinpointed the positions and intrinsic movements of more than 100,000 stars with a precision of up to 0.001 arc seconds. This enables the distances of stars within several thousand light years of us to be determined very precisely. In addition, more than a million stars were astrometrically surveyed with lower precision and their luminosities and colours were determined. Among other advantages, these measurements are critical for determining the distance scale. The precision achieved by Hipparcos will be substantially improved by GAIA, a future follow-up mission.

Astrometry – calibration of the distance scale and the stellar parameters

From interstellar clouds to protostars

The second component after stars is the interstellar medium (i.e., the gas and dust between the stars). New stars are still created today within the interstellar medium. Here gas condenses to form large clouds of up to one million solar masses. Smaller parts of these clouds can collapse under their own gravity and, in the course of this collapse, flatten to rapidly rotating protostellar disks. The matter at the centre of these disks then condenses to form stars.

Protostars identified

One of the problems that arise when studying the initial phase of the creation of stars is that it takes place hidden inside dense clouds. These clouds only becomes translucent at longer wavelengths that include the far infrared, submillimetre, millimetre and radio ranges. Using the German-French-Spanish millimetre-range observatory IRAM and the ISO satellite, it was possible for the first time to detect protostars, the extremely cold and dense zones in the interior of dust clouds. These "star cradles" will be studied in detail in the future using observatories such as SOFIA, Herschel and ALMA.

Protoplanetary disks

One very important advance was the first identification of protostellar dust disks using the IRAM telescope and the Hubble Space Telescope (Figure 1.4). These young stars are now in a similar phase to our sun 4.6 billion years ago. Further research must clarify how these disks formed and under what conditions planets were created within them. To pursue these questions, observations over all wavelength ranges, measurements from a variety of astrophysical laboratories, and numerical simulations are necessary. Infrared interferometry, which is to be carried out using the *Very Large Telescope* (VLT), the *Large Binocular Telescope* (LBT) and DARWIN, a European satellite project, will be of particular importance for future research of stellar and planetary genesis. Using the *James Webb Space Telescope* (JWST) it will be possible for the first time to resolve and optically observe young solar systems.

Jets from young stars

In the mid-1980s, it was also surprisingly realised that many young stars eject tightly bundled particle beams into space and perpendicular to the disk plane (Figure 1.5). These jets are similar in their morphology to the relativistic particle beams observed in active galaxies (see above), but are only up to ten light years long and have much slower particles. The details of their genesis remain unclear. The cause and energy source may lie within the accretion disk surrounding these young stars, similar to the gas jets of active galaxies. Another subject of current research is the question of how the approximately

1.2 The life cycle of stars and the matter cycle

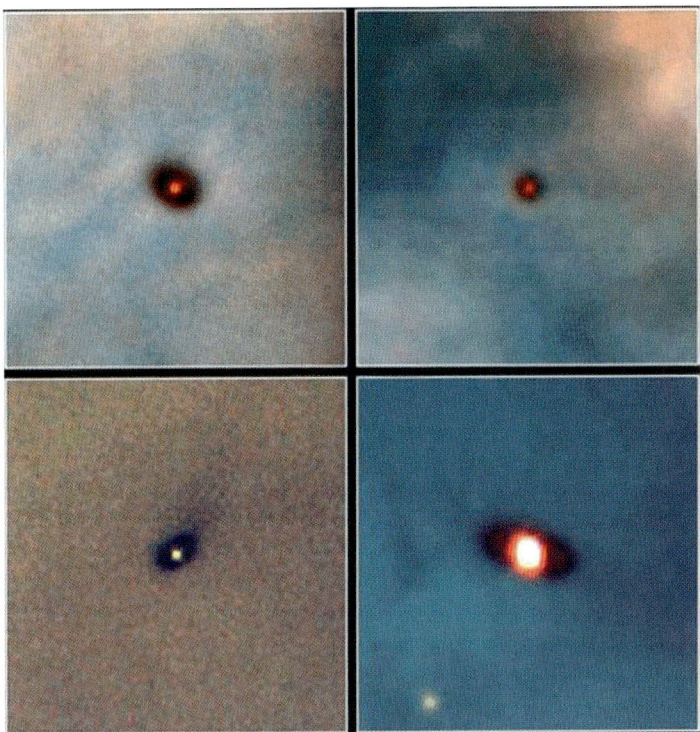

Fig. 1.4: Dark dust disks around young stars, discovered by the Hubble Space Telescope, made visible by absorption of the radiation from the lighter background. (NASA/ESA/MPIA/AIP)

ten-thousand year long jet phase affects the development of the star and the disk.

Extrasolar planets

The pioneering discovery in 1995 of planets orbiting nearby stars opened up a whole new field of astronomical research. More than a hundred of these extrasolar planets were known by early 2003. Currently, almost all of these bodies can only be indirectly detected by way of the gravitational effects on their stars. One of our foremost aims is to directly observe these objects within the next 10 to 15 years using imaging and astrometric techniques. The direct detection of massive, Jovian, extrasolar planets and the spectroscopic investigation of their atmospheres could soon be possible thanks to the improved adaptive optics on new large telescopes. Advances are also expected from GAIA, the astrometric space telescope.

Only planets at least as massive as Saturn can be detected using today's resources. Detection of earth-like planets and the identification of their chemical properties would represent

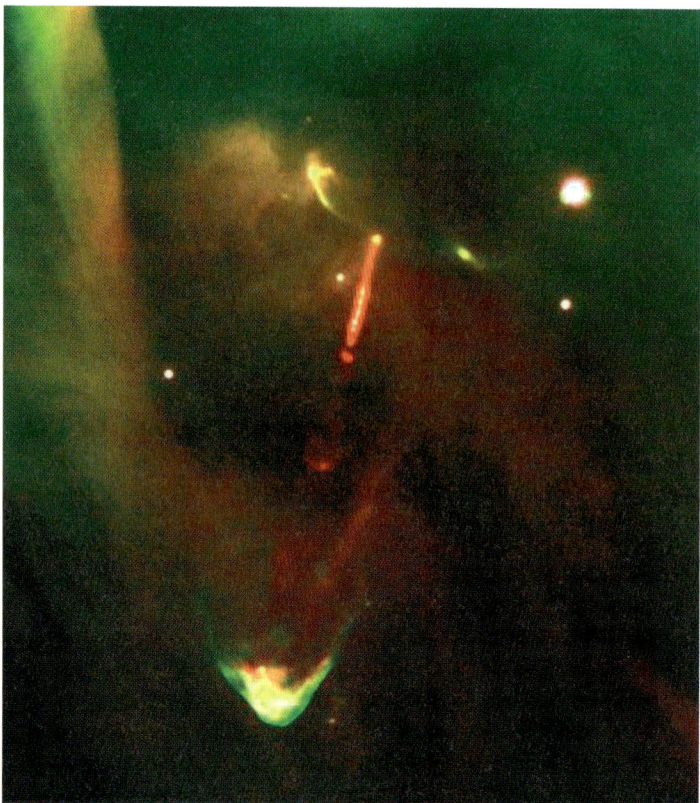

Fig. 1.5: A jet is emitted from a young star in the star formation region in Orion and ends in a frontal wave. (ESO/LSW Heidelberg)

an enormous move forwards. Extremely precise photometric methods (observations of occultation of the star by the planet) and direct imaging techniques (infrared or optical interferometry) using space observatories are necessary to achieve this very demanding objective. The requisite instruments, such as DARWIN, are currently being planned by researchers in Europe and the USA.

Our solar system as a model

At this point it is worth mentioning solar system research, which is, however, not the subject of this memorandum. A profusion of what are presumably generic characteristics of planetary genesis can be derived from new discoveries made in the field of cosmochemistry, in terms of the chemical and physical properties of "primordial matter" in our solar system. This information can be useful when searching for extrasolar planets. It is also necessary in order to understand whether or not our solar system is typical or occupies a special position among

1.2 The life cycle of stars and the matter cycle

the planetary systems. Researchers at German institutes have also made vital contributions in this field over the last 15 years. These include a theoretical link between the development of protostellar disks and measured cosmochemical variables, and the dating of primitive meteorites. Direct analyses of the composition of cometary material by the ESA probe Giotto was, without question, one of the highlights.

A star is illuminated by the nuclear reactions within its interior, which thereby release energy. It remains stable until it has consumed a considerable amount of its fuel reserves. The way in which it ends its existence depends on its mass. Our sun will expand to a red giant in about five billion years and then eject its outer gas shell. It then collapses to form a white dwarf the size of the Earth. The temperature increases to several ten-thousands of degrees and illuminates the previously ejected gas shell. These shells, known as planetary nebulae, are observed in numerous forms throughout the Milky Way and nearby galaxies (Figure 1.6).

Final stages of stellar development

Very massive stars, around 8 to 30 times more massive than the sun, first expand to supergiants and lose a great deal of their matter, before they explode as supernovae. While the outer shell is ejected, the innermost region of the star collapses upon itself forming a neutron star with a diameter of only around 20 kilometres. Neutron stars are objects with fascinating properties. The temperature of a young neutron star is approximately 100,000 degrees, as demonstrated by ROSAT. In the interior, matter is as heavily compressed as in an atomic nucleus. A piece of this matter the size of a sugar cube would weigh ten billion tonnes on Earth.

Neutron stars and pulsars

If a star collapses to form a neutron star, it also rotates extremely fast around its own axis. The neutron star in the Crab Nebula (Figure 1.7), for example, rotates around its axis 33 times per second. At the same time, the magnetic field at the surface becomes enormously dense and is finally a trillion times stronger than Earth's magnetic field. Electrically charged particles are accelerated almost to light speed along the axis of the magnetic field and discharged into space. They emit (synchrotron) radiation in the direction of discharge and the particle swarm generates two cones of light, which project into space from the magnetic north and south poles respectively. This radiation encompasses the radio range up to the highest gamma energies

In many neutron stars the axis of the magnetic field is inclined with regard to the axis of rotation. This causes the two

1 Astronomy yesterday, today and tomorrow

Fig. 1.6: The brightest planetary nebula in the northern sky: the Dumbbell Nebula, 650 light years away. (ESO)

light cones to rotate through space. If they by chance pass over the Earth, they can be detected by their lighthouse-like flashing pulse. These flashing neutron stars are called pulsars.

Pulsars have also become important because they represent extremely precise "clocks" due to their practically constant pulse rate. The American astrophysicists Taylor and Hülse utilised this property to measure the mutual orbits of two neutron stars. Over a number of years they realised that the orbital period was gradually slowing. It was possible to explain this effect by the emission of gravitation waves. Bodies lose energy in this way and slowly approach each other. The measured reduction corresponded to less than one percent, with the energy loss predicted by the General Theory of Relativity. The two researchers were awarded the Nobel Prize for Physics in 1993 for this first indirect proof of gravitational waves.

Stellar black holes An extremely massive star with more than 30 solar masses can probably not avoid collapse. In theory, the burned out

1.2 The life cycle of stars and the matter cycle

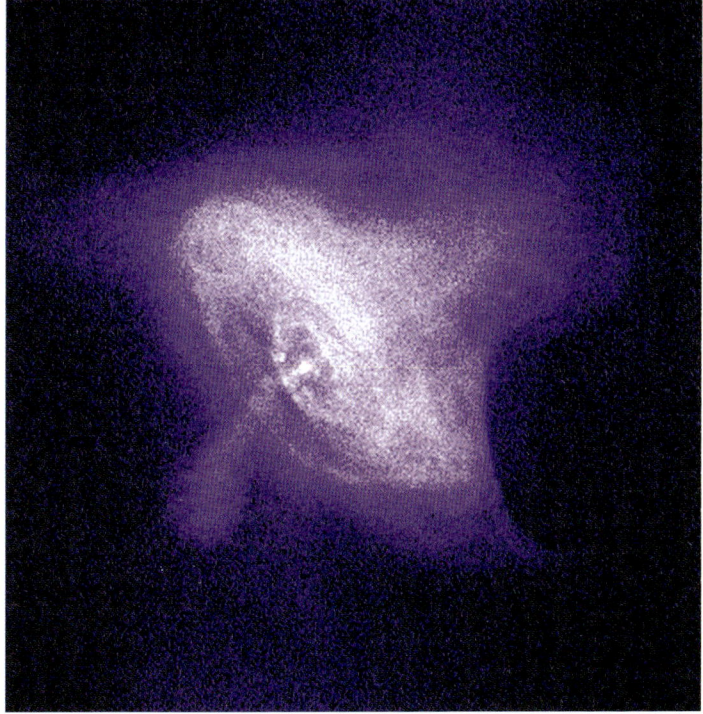

Fig. 1.7: The crab supernova remnant in the constellation Taurus. The supernova was observed by Asian astronomers in 1054. Top, image in the visible band (ESO), below, the X-ray image taken by Chandra (NASA/ CXC/ SAO), showing details of the inner, active region around the quickly rotating neutron star.

star collapses completely and disappears in a (stellar) black hole. The gravity is now so strong within a certain region that neither matter nor light can escape. The radius of this region in a black hole of eight solar masses, for example, is about 25 kilometres.

Black holes can only be detected indirectly due to the effects they have on their environment, as exemplified in double star systems, where two stars orbit one another in gravitationally mutual orbits. Astrophysicists today are aware of several systems in which one component is invisible and very massive, therefore probably being a black hole. X-ray observatories, such as the European XMM-Newton space telescope and future XEUS, are extremely important in this field. Interferometers on the VLT and LBT, currently under construction, will also provide more information on these mysterious heavenly bodies.

The matter cycle

In around five billion years from now, our sun will expand to form what is known as a red giant. Internal processes initiate an expansion of the outermost gas shell, simultaneously leading to its cooling and forcing the radiation peak into the long-wave, red range. A cycle, or matter "recycling," is associated with the birth and demise of the stars; stars are re-created from the interstellar medium. During their stable combustion phase they generate by nuclear fusion heavy elements in their interiors. As red giants or supernovae they then give back some of this processed matter to the surrounding interstellar gas, where it serves as the raw material for the next generation of stars. In this way the stars gradually enrich the interstellar medium and subsequent star populations with heavy elements.

Cosmic evolution of the elements

This process represents one of the prerequisites for the emergence of the planets and, ultimately, life itself. This is because, as far as we know today, only the light elements hydrogen and helium were created during the big bang. The diversity of chemical elements and all life on our planet is therefore a result of a cosmic chemical process that took place long before the Earth existed. Every carbon or oxygen atom in our bodies was generated billions of years ago in the interior of a star.

Supernova remnants

Understanding of these complex nucleosynthetic processes (i.e. the formation of heavy elements), is crucial and possible by observing planetary nebulae and the remnants of supernova explosions. The latter are generally extremely hot and must therefore be investigated in the X-ray range (or in the nonthermal radio or gamma ray range). Many new super-

1.2 The life cycle of stars and the matter cycle

nova explosive clouds have been discovered with the help of ROSAT. However, spectral investigations of the requisite sensitivity and resolution have only now been made possible thanks to the ESA XMM-Newton mission and the American X-ray observatory Chandra.

Supernova remnants are also seen as the principal source of high-energy charged particles, known as cosmic rays. They fill the entire Milky Way, and probably intergalactic space, and bombard the Earth constantly. They are considered one of the motors of the natural gene mutations that life is subjected to on our planet. The discovery and development of what is probably the most important process for the generation of cosmic rays is based on the research results acquired by German institutes.

The numerical simulations of supernovae also need to progress in parallel with observations. The explosive mechanism is still not completely understood. Neutrinos and chaotic mixing processes during detonation presumably play

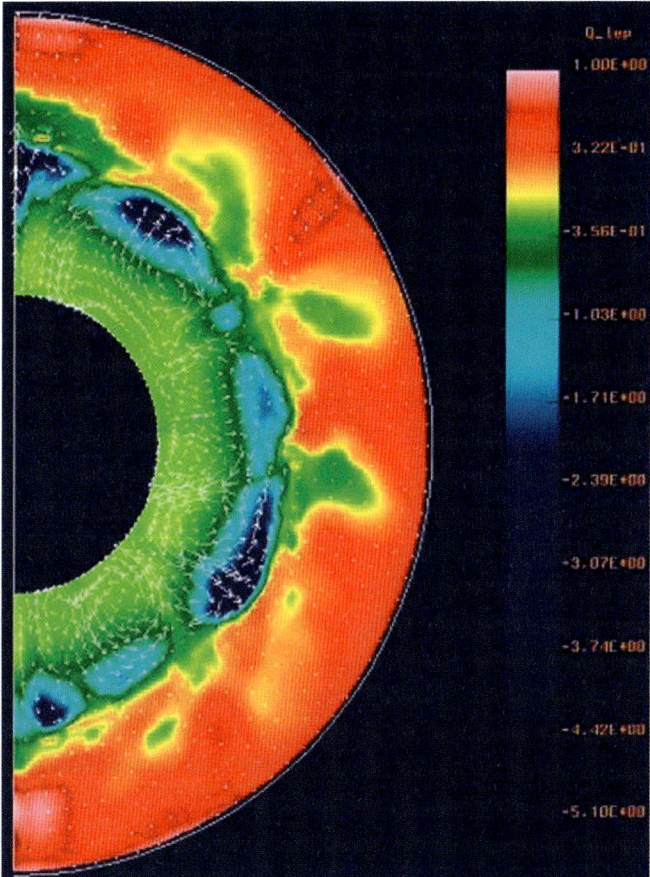

Fig. 1.8: The explosion of massive stars can be simulated on the computer. The turbulent mixing of matter can be seen: oxygen, blue; silicon, green; nickel, red. (MPA)

Our sun

a decisive role. One of the objectives of such investigations is to identify the frequency with which the various chemical elements, up to and including uranium, are generated (Figure 1.8).

Our nearest star is the sun. Due to its proximity, it allows plasma physics processes of fundamental astrophysical importance to be investigated with the necessary spatial resolution. The magnetic activity of the sun leads to fluctuations in its radiation and in the solar wind. They both influence satellites, communications systems and probably our climate. In past years, space telescopes such as SOHO and terrestrial telescopes such as the observatory on Tenerife have analysed the sun with increasing precision (Figure 1.9). One of the highlights of this research was the precise analysis of the inner structure of the sun with the help of the relatively new method of helioseismology. Huge advances have also been possible in high-resolution imaging and spectroscopy (plasmadiagnostics) of the solar atmosphere, including the corona, using instruments in the visible, ultraviolet and X-ray radiation ranges. Two highlights of German solar research include the precise identification of the sources of solar wind using SOHO and the discovery that the solar magnetic field has strengthened manifold, apace with global warming, since the Little Ice Age of the seventeenth century.

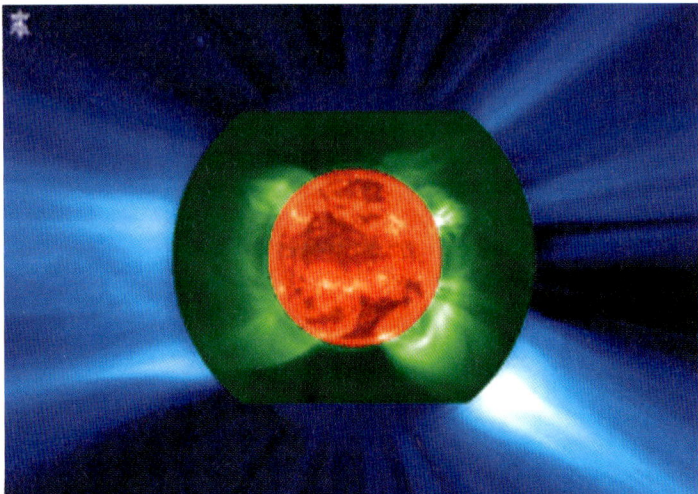

Fig. 1.9: Solar observations using various instruments on the SOHO satellite. This image shows areas of magnetic activity on the red surface of the sun, the hot solar corona (green) and the solar wind (blue). (ESA/NASA/MPAe)

Holistic studies of the sun and its atmosphere, extending into interplanetary space, will form a focal point of future research. This "holistic" view is the key to understanding magnetic activity and the influence of the sun on the Earth's climate. The necessary wavelength coverage and spatial and temporal resolution will be achieved by the next generation of telescopes, including GREGOR on Tenerife and the balloon-based SUNRISE, through possible participation in the American solar telescope project, ATST, and with the instruments on the ESA space probe SOLAR ORBITER.

1.3 New windows into space

Progress in astronomy and astrophysics has most often been associated with the development of new instruments, detectors and telescopes, both in terms of the increased performance of the instruments in the "classical" wavelength ranges and in opening up new fields. In addition, this often leads to very fruitful synergies with developments in totally different fields of physics and technology, such as semiconductor and laser physics, electronics and computer technology. Technological progress provides new information by achieving better spatial resolution and more detailed analyses of the spectrum of electromagnetic waves.

Pioneering work in opening new windows into space was recognised by the Nobel Prize for Physics in 2002. It was awarded jointly to Riccardo Giacconi (1962, 50%) for the discovery of cosmic X-ray radiation, Raymond Davis Jr. (1971, 25%) for the solution of the solar neutrino problem, and Masatoshi Koshiba (1987, 25%) for the discovery of the first astrophysical neutrinos from supernova 1987A. During the next 15 years, astrono-mers want to further open windows to the neutrino universe and also a new window to gravitational waves.

For example, the sensitivity of radio telescopes has doubled on average every two to three years since the nineteen-thirties (Figure 1.10). This rapid development is easily comparable to Moore's (semiconductor) Law, stating that the storage capacity of chips doubles every $1^1/_2$ years. Today's radio telescopes equipped with the newest receivers are around one hundred million times more sensitive than 50 years ago. More technical advances should lead to sensitivity increases of between 10 and 50 times within the next 10 to 15 years. These improvements, already demonstrable today, open up fantastic oppor-

"Moore's Law" applied to the sensitivity of telescopes

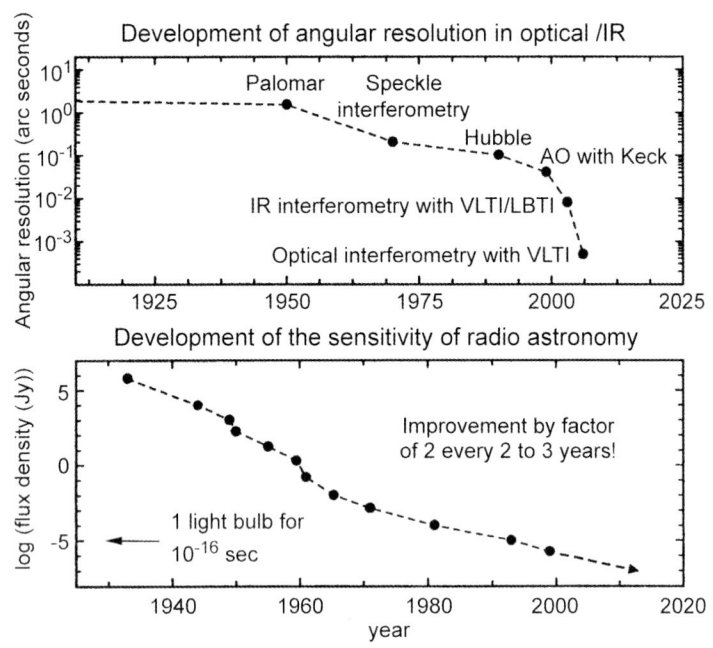

Fig. 1.10: Development of the measuring sensitivity in radio astronomy (bottom) and angular resolution (top) of telescopes in the infrared and visible light ranges. (MPIfR)

tunities for astronomy in the submillimetre and millimetre ranges. In the future, increases in the speed of digital signal processing will also set in motion a completely new class of digital radio telescope in the centimetre range.

Increase in angular resolution

Angular resolution (the capacity to discern two close objects as separate entities) has also improved considerably in the past (Figure 1.11). Here, radio astronomy occupies a special position inasmuch as the idea of coupling two or more antennas (synthetic aperture) had already been developed by the mid 1950s. From the end of the 1960s this resulted first in intercontinental interferometry and then in space-based interferometry, which was successfully demonstrated at the end of the 1990s. It allows an angular resolution of up to 50 millionths of an arc second, corresponding to a width of around 10 centimetres at the distance of the moon!

Optical and infrared interferometry: a promising field

Building interferometers in the infrared and visible light ranges is a much more difficult undertaking. Initial successes with the Keck Observatory in Hawaii and the Very Large Telescope at the European Southern Observatory, ESO, in Chile, promise great leaps in angular resolution in the near future.

1.3 New windows into space

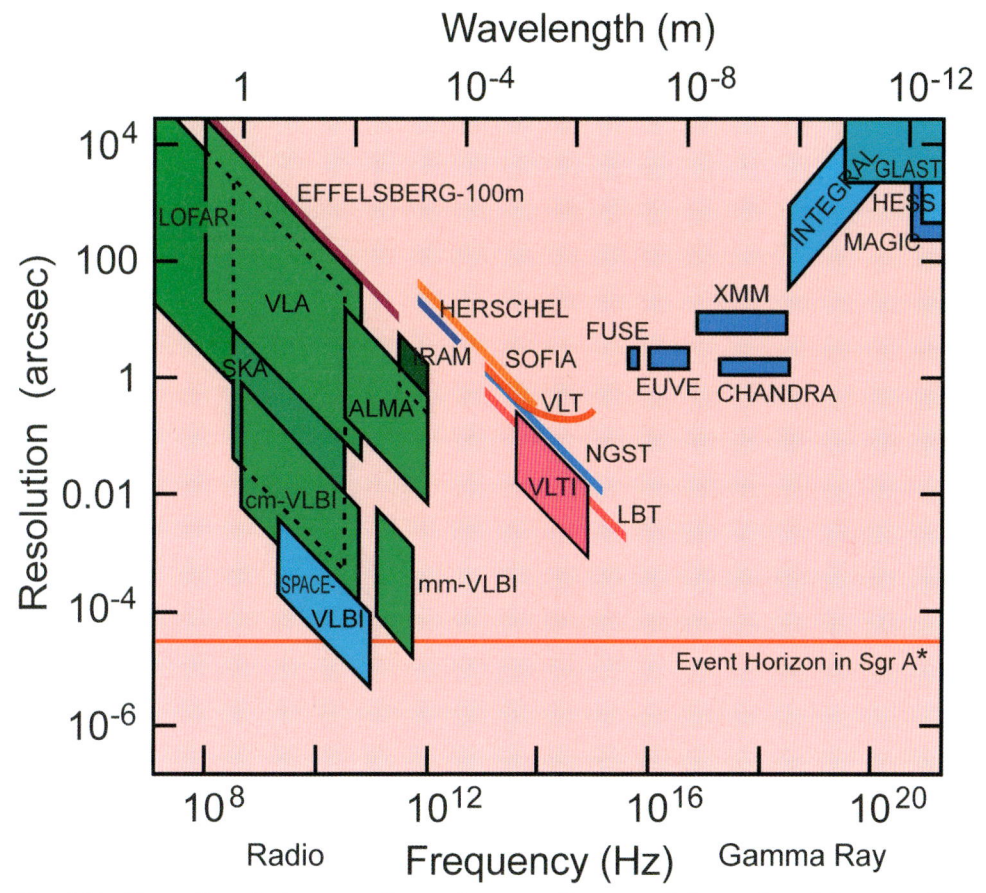

Fig. 1.11: The entire electromagnetic spectrum can be covered by modern telescopes. The angular resolution and sensitivity range of the most important instruments mentioned in this memorandum are shown here.

Adaptive optics

Adaptive optics technology plays a vital role in the implementation of these projects. It allows natural air turbulence, which leads to blurring in astronomical exposures, to be compensated for during observation. Adaptive optics, as also developed by German groups for Calar Alto and for Tenerife, operate very well in the infrared. However, in order to extend the technology into the visible light range, additional effort is still required. Interferometers, equipped with adaptive optics, will improve angular resolution more than one hundredfold in the coming 10 to 15 years. In other wavelengths, as well, new technologies and telescopes will lead to improvements in angular resolution.

Neutrino astrophysics	Neutrinos are elementary particles that only interact weakly with normal matter, making them very difficult to detect. They are created during nuclear fusion in the interiors of stars or in supernova explosions. These particles are therefore very important to astrophysics, because they contain unique information on the processes discussed here and allow us, for example, to take a look into the interiors of stars, a view that would otherwise remain hidden.
Solar neutrinos	Since the end of the 1960s, several experiments have been carried out in an attempt to locate solar neutrinos. The detection of solar neutrinos and the discovery of neutrinos from supernova 1987A were honoured with shares of the 2002 Nobel Prize for Physics. At the beginning of the 1990s several experiments, including the German-led GALLEX, caused sensations when they managed to detect neutrinos in the main branch of the hydrogen fusion cycle. Strictly speaking, it was the first direct evidence that the sun – and therefore all other stars – generates its energy by nuclear fusion. However, the detectors registered significantly fewer neutrinos than theory had led us to expect, which was explained by the fact that neutrinos possess an at-rest mass. This contradicts the standard particle physics model, which states that neutrinos are massless. These observations are merely one example of the many interactions of astrophysics with other sub-disciplines of physics.
Neutrino research in ice	Another success was the detection of a dozen neutrinos from supernova 1987A, located in the Large Magellanic Cloud. Here, the Japanese detector Super-Kamiokande enjoyed particular success. Encouraged by these initial successes, new "neutrino observatories" were planned and built. The most progress so far has been made by the AMANDA project, in which German researchers play a decisive role. Light-sensitive detectors are sunken into the ice near the Amundsen-Scott Station at the south pole. Similar to Super-Kamiokande, they register energetic neutrinos, which emit Cherenkov radiation inside the detector. Researchers hope to actually detect neutrinos from supernovae or from the centres of distant galaxies using AMANDA. Even higher sensitivity will be achieved by the international follow-up experiment IceCube; similar to AMANDA, it will be built below the south polar station.
Gravitation waves	Astrophysicists around the world broke new technological and scientific ground with the construction of gravitational wave detectors. They will make it possible to directly detect temporally variable ripples in space, as predicted by the General

Theory of Relativity. The challenge lies in the fact that the expected change in spacing amounts to a mere tiny fraction of an atomic nucleus. It is hoped to detect this by using laser interferometers. These instruments have been made possible by advances in the field of laser and precision metrology. German researchers are in charge of the construction of the GEO600 interferometer near Hanover in northern Germany.

Proof of this phenomenon would represent a further brilliant confirmation of Einstein's General Theory of Relativity. On the other hand, the waves would provide information on the most energetic processes in the cosmos. These include supernovae, binary star systems following close orbits, colliding and merging neutron stars, and black holes. It is even hoped to obtain gravitation signals from the big bang in the long term. But this will require laser interferometers in space, such as LISA, as currently planned by ESA and NASA.

Besides classical theoretical astrophysics, during the last few years numerical simulations have become an important aid for describing complex phenomena in our universe. The performance of computers has increased to the same degree that telescopes and detectors have advanced. During the last 15 years the speed of computers has increased by a factor of around one thousand. In conjunction with the development of increasingly efficient algorithms and the use of specific hardware components, this has resulted in enormous advances in numerical simulations for many astrophysical processes. These include the genesis of stars and planets in dust clouds, supernovae explosions, the genesis of galaxy clusters, or the processes in the disks around black holes, to name but a few examples.

Advances in numerical astrophysics

Realistic results are only obtained from numerical simulations if, on one hand, as many physical processes as possible are taken into consideration and, on the other, good spatial and temporal resolution is achieved. In order to interpret the ever more detailed observation data, increasingly powerful computers must be employed and the numerical methods must be continuously improved.

Box 1.1: Multi-wavelength ground- and space-based astronomy

The use of telescopes in all wavelength ranges, from radio to gamma waves, which correspond to ten to the power of minus 15, is increasingly important to modern observational astronomy. There are a number of objects whose nature only becomes obvious when seen in all wavelengths, active galaxies being one very good example. Analysis and interpretation of the data requires increasingly complex software, which will be made available to all users within the framework of the "Virtual Observatory" project.

Since the beginning of the 1980s, multi-wavelength astronomy has been subject to rapid development and German groups have played a leading role (Figure 1.11). The German-French-Spanish institute IRAM (Institute for Radio Astronomy in the Millimetre Range) has provided German researchers with access to the world's best telescopes in the millimetre wavelength range for more than 10 years. During the 1990s the European Space Agency's (ESA) *Infrared Space Observatory* (ISO) represented a further highlight. It was the world's most successful infrared mission of the 1990s. German institutes and industry were crucially involved in building the detector, contributing to the astronomical yield and to the data archive.

The X-ray space telescope ROSAT was also a great success. It carried out the first complete survey of the sky using an imaging X-ray telescope and discovered more than 100,000 individual sources. To date, approximately 5,000 scientific publications have been made based on ROSAT data. This makes this telescope almost as successful, in terms of data yield, as the Hubble space telescope. In the meantime, the follow-up missions XMM-Newton and CHANDRA provide considerably more detailed information on the X-ray universe. A milestone of gamma ray astronomy was NASA's *Compton Gamma Ray Observatory* (CGRO). It carried four scientific instruments, one of them designed under German leadership. Compton allowed the observation of particularly hot heavenly bodies such as neutron stars or supernovae explosion clouds with previously unachievable sensitivity.

2 The scientific issues

This chapter presents examples of some of the most current and exciting research topics in modern astronomy and astrophysics. On the one hand, recent research and future development prospects in these fields were taken into consideration when selecting the core topics and associated materials. But, on the other hand, the highlights and strengths of German research also needed to be taken into account. The presentation of research fields given in this chapter is not exhaustive; rather, it is intended to give an impression of the many fields open to modern astrophysics and to future developments. These topics are then cited as examples in Chapter 3, where the enormous scientific and technical advances in observatories and instruments expected to come about in the next 15 years are discussed.

2.1 The universe – its origin, evolution and large-scale structure

Scientific advances

- Discovery of structures (anisotropies) in the cosmic background radiation.
- Determination of the geometry of the universe.
- Evidence of dark matter and vacuum energy.
- Recalibration of the cosmic distance scale.
- Dating of the oldest stars in the Milky Way and the universe.
- Characterisation of the large-scale structure of the universe.
- Discovery of gravity lenses and their application as astrophysical "instruments".

2 The scientific issues

- Description of the structures in intergalactic gas and its enrichment with heavy elements.

Purpose and aims

- To precisely determine the geometry (curvature) of the universe and the governing cosmological parameters.
- To precisely calibrate the distance scale.
- To reconstruct cosmic history: the origin of the first stars, the galaxies and galaxy clusters.
- To reconstruct the chemical history of the universe: production of heavy elements.
- To identify the large-scale distribution of luminous matter (stars and gases) and dark matter.
- To detect direct evidence of the constituents of dark matter.
- To discover gravitational waves.

The recent history of cosmology

As a result of the rapid advances made in observational methods, in conjunction with theoretical developments, cosmology, the science of the history of our universe, has developed from what was originally a predominantly mathematical tradition to represent a central discipline of astrophysical research. As late as the 1980s, cosmology was chiefly shaped by purely theoretical developments such as the theory of the inflationary universe. It was also at this time that the first attempts were made to simulate the origin of large-scale structures with the aid of computers.

Today, it is possible to study far distant galaxies in the early universe by direct observation. This is possible because, due to the finite travelling time of light, every look into the depths of space also represents a look into the distant past. It has only been possible for around ten years to look so deeply into the universe that the evolution of galaxies can actually be observed. The most distant sky bodies emitted the light detected today at a time when the cosmos was still at a very early stage of development. In addition, large-scale examinations of the sky to determine the spatial distribution of millions of galaxies have been operating for the last five years. This will allow the structure of today's universe to be directly surveyed.

Moreover, investigation of the microwave background and its anisotropies provides us with information on the structure of the universe as it was only 380,000 years after the big bang. These observations, carried out during the last ten years, have endowed us with fundamental information about the history of the universe. Together with other observations,

2.1 The universe – its origin, evolution and large-scale structure

these results have led to a self-consistent standard model of the cosmos, characterized on the one hand by the big bang model, in conjunction with inflation theory, and by the matter content of the universe on the other. Only around 4% of the matter forming the universe consists of the chemical elements of "normal", baryonic matter. There are also two fundamentally new forms of matter: dark matter and vacuum energy, which can both be detected only by their gravitational effects. The presence of a vacuum energy component, around 70% of the total mass density, drives the acceleration in the universal expansion.

If these results are confirmed, they would indicate that our current understanding of elementary particle physics is incomplete, because the origin of dark matter and vacuum energy must lie in microphysics. Clarification of the physical nature of these types of matter represents an enormous challenge to physics and will lead to a revolution in our understanding of the fundamental laws of nature. The principle quantities for describing the universe, the cosmological parameters, were recently determined with a precision of a few percent by the combination of several observations, heralding the era of precision cosmology. This field will probably reach full flower during the next decade.

2.1.1 The big bang and cosmic background radiation

The big bang

In the big bang model, which today is generally regarded as confirmed, the physical laws of nature governing elementary particles meet the space-time theory of the general theory of relativity. The universe thus becomes a unique laboratory, far beyond anything possible in an earth-bound experiment, and provides the motivation and testing opportunities for the unification of all physical laws.

After the big bang the expanding universe began to cool. First, the elementary particles formed in the primordial gas and, after about three minutes, the lightest atomic nuclei took form. Electrons and nuclei first formed a plasma opaque to light rays. It was not until around 380,000 years after the big bang that the universe had cooled enough to allow electrons and nuclei to unite to form atoms. The universe then became transparent and radiation was able to propagate practically unhindered. Due to the continued expansion of the universe, the radiation lost energy and cooled to today's 2.73 Kelvin, with an intensity maximum in the microwave range. The discovery of the omnipresent cosmic microwave background (cosmic background radiation) by Penzias and Wilson in 1965, predicted in 1946 by Gamov, is one of the greatest triumphs of cosmology and forms one of the principal pillars supporting the big bang theory.

The seeds of structure formation

According to theory, the galaxies and galaxy clusters formed from minor density fluctuations in the primordial gas. These "seeds of the earliest structures" are said to have left their mark in the cosmic background radiation. Sensationally, the American COBE satellite was able to discover these predicted anisotropies in 1992 with relative amplitudes of around one thousandth of a percent (Figure 1.1). This represented an important initial confirmation of the theory of structural genesis in the universe.

However, COBE had relatively poor vision. The smallest recognisable structures in the sky were about 7° wide. Copious amounts of information on the structure of the universe at an age of 380,000 years are hidden in the cosmic background radiation at angular scales between around one degree and several arc minutes. New and more powerful experiments will therefore measure these fluctuations with vastly improved angular resolutions and greater sensitivity. Two satellite experiments are intended to fulfil these tasks: NASA's *Wilkinson Microwave Anisotropy Probe* (WMAP), started in 2001, and the European Planck mission, which will launch in 2007 (Figure 2.1). A preview of the spectacular results expected was given by the microwave mapping of small regions of the sky with improved angular resolution, by the balloon-based instruments BOOMERanG and MAXIMA in 2000, and by the recently published WMAP map. They demonstrated that the universe is "flat". This means that the mean (energy) density in the universe corresponds precisely to the critical density, in agreement with the predictions of the theory of the inflationary universe.

Planck will produce sky maps around 50 times more detailed and around 10 times more precise than COBE and has an angular resolution three times better and much broader wavelength coverage in comparison to WMAP. German astronomy participates in the evaluation of the Planck mission primarily by providing a data centre.

The end of the "cosmic Middle Ages"

Once electrons and atomic nuclei were united to form atoms, the universe became transparent to relatively "long-wave" radiation. This did not apply to photons in the ultra-violet part of the spectrum, which are very efficiently absorbed by atoms, explaining why this epoch is known as the "dark age".

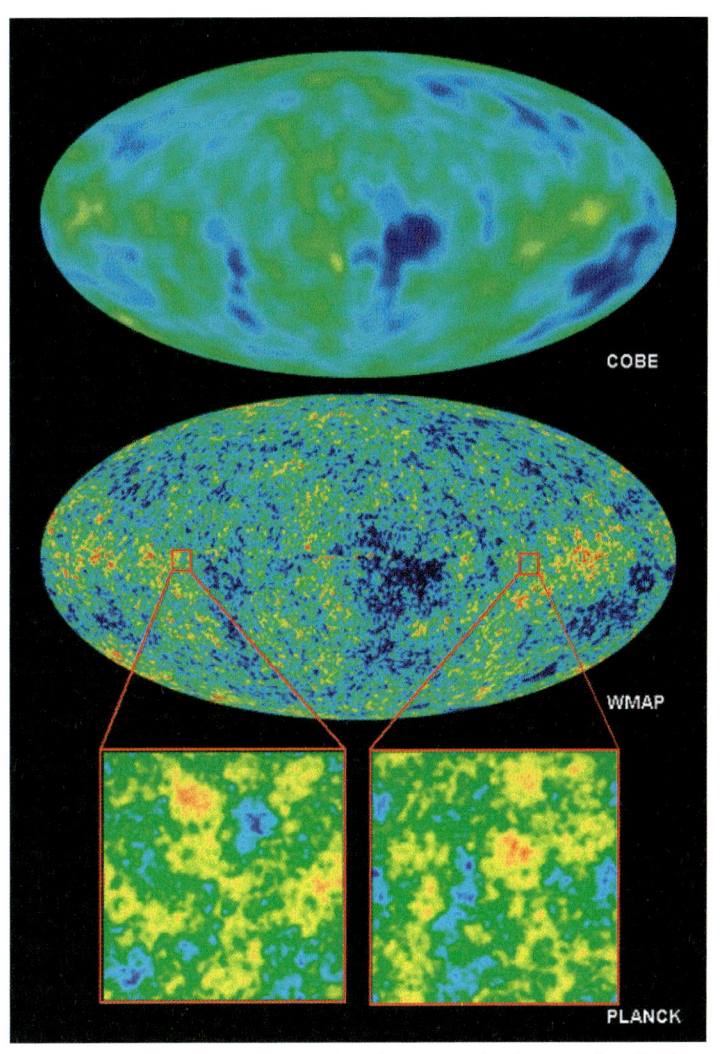

2.1 The universe – its origin, evolution and large-scale structure

Fig. 2.1: Structures in the cosmic microwave background radiation. Whilst only a rough picture of early cosmic structures was achieved by the COBE satellite (top), NASA's Wilkinson Microwave Anisotropy Probe (WMAP) surveyed the expected quasi-periodic spatial fluctuations (centre). ESA's Planck satellite will be capable of recording the finer structures in exquisite detail (bottom). They contain the key to the fundamental cosmological parameters and the physical conditions prevalent shortly following the big bang. (Source: NASA/WMAP Science Team/MPA)

However, because we can observe quasars with very great redshifts, the universe as a whole must have been re-ionised in the meantime. The high-energy photons required probably originated in a first generation of luminous sources such as the first quasars or galaxies. The search for these sources, and identification of the cosmic epoch in which they became incandescent, forms a fascinating aspect of observational cosmology and will provide decisive information on the physical conditions at the time these objects were created. The complex structures expected at the transition from a neutral to a primarily ionised universe are the subject of intense theoretical research and may, in the future, be directly observed by the *Square Kilometre Array* (SKA).

2.1.2 The cosmic distance scale

Edwin Hubble's discovery that all galaxies are moving away from each other was a groundbreaking piece of modern research and revolutionized the beliefs of mankind. He discovered that the "recession velocity" increases as the distance of the galaxies from the Milky Way increases. This allows the distance to be calculated from the measurable relative velocity, if the proportionality factor, the Hubble constant, is known. At the same time, this constant determines the *scale* and *age* of our universe, whilst the density of the various states of matter describes the *shape* of the universe.

The relative velocity of a source is determined from what is known as the redshift of the spectrum, that is, from the measurable shift of characteristic spectral lines to longer wavelengths. Hubble's Law states that sources with large redshifts, that is with large relative velocities, are located a great distance from us. The radiation from these sources originated at a time when the universe was much younger; the redshift z is therefore not only a measure of the distance of a given source, but also of the corresponding age of the universe; Figure 2.2 demonstrates this relationship.

Calibrating the distance scale

Measuring distances in the universe has always represented one of the central problems of astronomy and cosmology. In order to determine the Hubble constant, methods must be found for establishing the distances of galaxies independently of their relative velocities. This calibration of the distance scale is one of the central tasks facing current research. It is done by way of "standard candles". These are usually star types of known luminosity. The value of this luminosity can be deter-

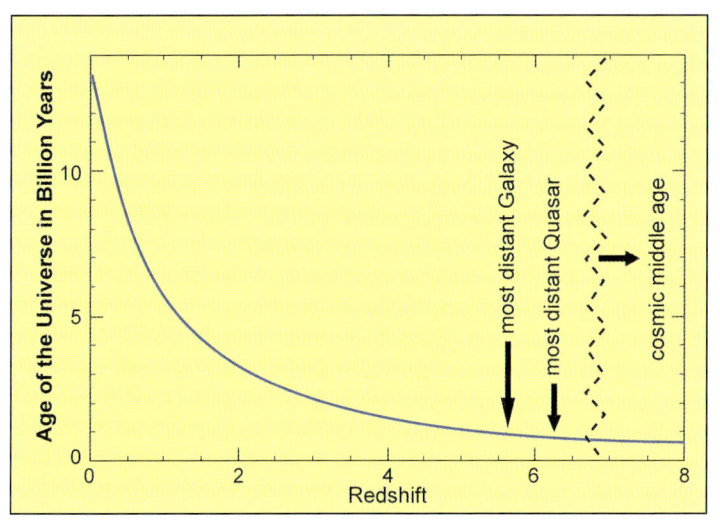

2.1 The universe – its origin, evolution and large-scale structure

Fig. 2.2: Relationship between the redshift of a source and the age of the universe, and the time at which the light was emitted by the source. The end of the "cosmic Middle Ages" is not exactly known. (Bonn University)

mined from the magnitude and the distance, using geometrical parallax methods, of a random sample of relatively close representatives of these star types. This method of calibration of the extragalactic distance scale has made important advances thanks to the European HIPPARCOS satellite. Very important in this regard is the future astrometry mission GAIA, which will determine the distances of a sufficiently large number of standard candles with considerably greater sensitivity and precision than HIPPARCOS.

Exact calibration of standard candles is now even possible by observing extragalactic sources, because today it is possible to measure geometrical distances up to several million light years (Figure 2.3) using intercontinental radio interferometry (VLBI).

Recently, Type Ia supernovae have also been utilised as standard candles (Figure 2.3). They are particularly interesting because they can be observed over greater distances than other standard candles (currently up to redshifts of around 2), due to their enormous luminosity. In recent years these observations have provided spectacular proof, together with measurements of the background radiation and of galaxy clusters, that the expansion of the universe is not being braked or decelerated but is in fact being accelerated. This acceleration is described in theory by the cosmological constant originally introduced,

Accelerated expansion of the universe

2 The scientific issues

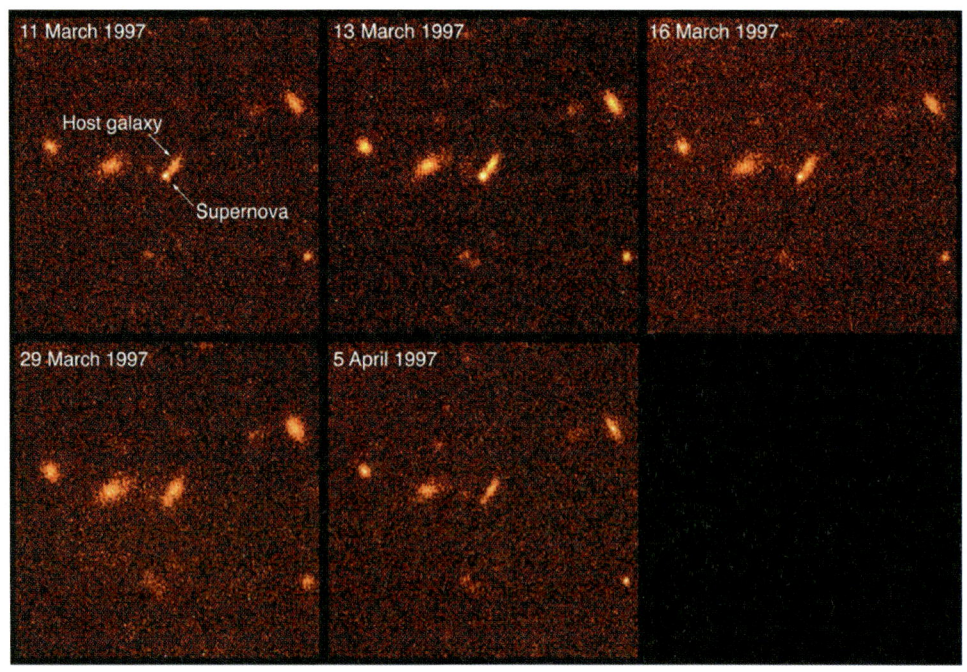

Fig. 2.3: Discovery of a supernova in a remote galaxy (z = 0.51) using ESO's New Technology Telescope. The object in the centre of the respective images first becomes brighter than the host galaxy and then fainter again during the course of the following month. (ESO)

and later rejected, by Einstein; the modern interpretation of this constant is represented by the previously mentioned vacuum energy. It is one of the most important, and simultaneously one of the most mysterious, parameters used for the description of cosmological models. In order to place the use of this relatively new method of distance determination using Type Ia supernovae on a firm foundation, further effort is required to better understand the physics of these exploding stars. Here, German theoreticians are making internationally relevant contributions.

The age of the universe

On the one hand, the age of the universe can be derived from the expansion law and the corresponding value of the Hubble constant. On the other hand, regardless of this, the oldest observable stars define a lower limit for the age of the universe. All attempts at dating are based on comparatively well known data originating in nuclear physics, which controls the evolution of stars. In contrast to the distance scale, dating the oldest stars in the Milky Way is based on low-mass, faint and cool objects. These may be located in globular clusters or exist as individual stars.

2.1 The universe – its origin, evolution and large-scale structure

Based on the data provided by HIPPARCOS, together with the most recent stellar evolution models, partly developed in Germany, the age of the oldest galactic globular clusters has been newly established during recent years. The new value of approx. 13 ± 3 billion years coincides well with the currently accepted cosmological measurement of the age of the Universe. GAIA will further improve on these observations of globular clusters and place them on a solid foundation.

Gravitational lenses

One of the most fascinating discoveries of the past 20 years is that of gravitational lenses. This gravitational deflection effect is described by Einstein's general theory of relativity and states that matter bends the surrounding space. If a beam of light enters an area of space which is influenced by the gravitational field of a galaxy or an entire cluster of galaxies, it must follow the resulting curvature of space and therefore travels along a curved path. Or in other words: light beams are deflected in a gravitational field in just the same way as massive bodies. If one observes a distant galaxy whose light has passed through the gravitational field of a foreground galaxy, the more distant object appears as a distorted arc or as a multiple image. The first double image of a quasar observed through a galaxy was discovered in 1979 and in the mid 1980s images of a galaxy which distorted a galaxy cluster to a ring shape were observed (Figure 2.4). Multiple images of radio quasars, which can be investigated at extremely small angular resolutions using

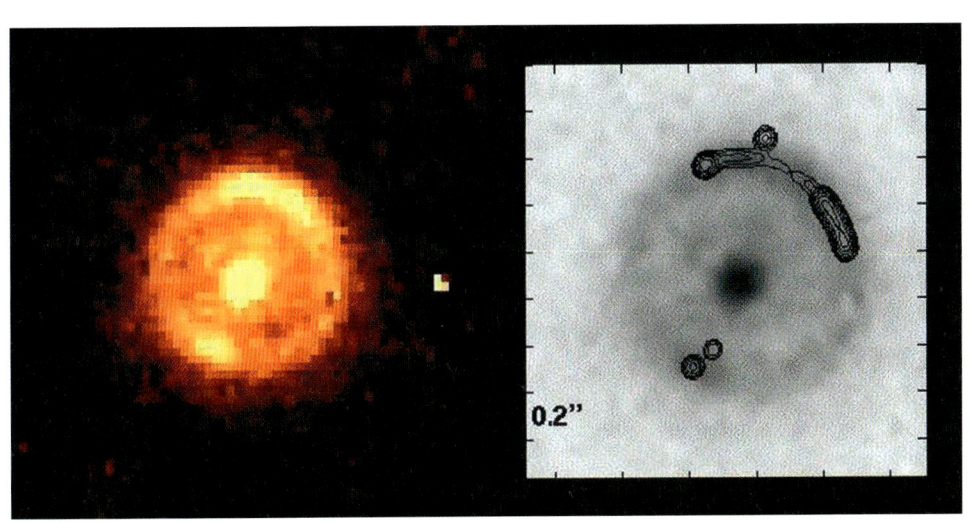

Fig. 2.4: In the gravitational lens system 1938+666 the remote radio galaxy is pictured as a ring ("Einstein ring"), to be seen in the HST/NICMOS image (left), whilst the associated radio source (right) is split into multiple images. (NASA/ STScI/Jodrell Bank/Bonn University)

Independent determination of the Hubble constant

interferometry techniques, are particularly informative. German scientists have made some very important contributions to this field.

Theoreticians have recognised that the gravitational lens effect can be utilised in a number of ways. For identifying the Hubble constant, for example, beside the very precise determination of masses. Fluctuations in magnitude must be traced in the multiple images of quasars, even though they cannot be viewed simultaneously because of the travelling time differences for the different light paths. Measurements of this type using optical and radio telescopes have already been carried out using a number of lens systems and have provided values for the Hubble constant which are located at the lower end of the scale estimated using other methods. Automated telescopes are being increasingly utilised to monitor magnitude fluctuations.

The difficulty in principle of using these methods lies primarily in the precise determination of the mass distribution of the lens galaxy and its environs. Its advantage, however, is that it is completely independent of the "classical" methods. Improvements to the lens models and a subsequent improvement in the precision of the Hubble constant are expected from future observations using the JWST.

2.1.3 Cosmological evolution

Understanding the evolution of the universe, from the early state that can be observed in the microwave background radiation, to the universe today, represents one of the central objectives of current cosmological research. Today, for the first time, it is possible to investigate and understand this evolution in detail and to thus directly witness a unique experiment in physics: our cosmos.

These new opportunities are being opened up by the development of new large telescopes in practically all wavelength ranges: sources with large redshifts, that is, in the early stages of the universe, are very faint due to their extreme distance, so they can only be investigated using very sensitive telescopes. The new large telescopes VLT, LBT and JWST are of primary importance to German astronomy for observations in the optical and near-infrared ranges. Observations in the far-infrared, submillimetre and X-ray ranges will be of primary importance for studying the creation phase of those galaxies displaying intense stellar production and presumably with

2.1 The universe – its origin, evolution and large-scale structure

very high levels of dust. Here, Herschel, APEX and ALMA, XMM-Newton, and later XEUS, will be employed.

It is essential that new theoretical approaches be developed in parallel to the quantitative interpretation of the collected data. This is facilitated on the one hand by understanding the physical processes that play a governing role in cosmic evolution, and in a qualitative understanding of the central processes involved. On the other hand, the complexity with which structures develop demands that these developments are re-enacted using what are, on occasion, highly complex simulations on very powerful computers; a quantitative comparison of these simulations to actual observations then leads to an awareness of cosmic evolution.

The large-scale structures in the matter distributed throughout the universe were formed from the interaction of cosmic expansion and gravitational forces that are represented by the visible fluctuations in the microwave background of the early universe.

Formation of structures in the universe

This evolution is dominated by dark matter, which only interacts gravitationally. Its evolution can now be traced in amazing detail by means of simulations (Figure 2.5). German institutes play a world-leading role in the design, implementation and theoretical interpretation of these analyses. They indicate that the earliest existing massive objects have relatively low masses, whilst the large galaxies and galaxy clusters only formed later, among other means by the fusion of smaller objects. This hierarchical model of the structural evolution of the universe has now been confirmed by numerous observations, for example by the increasing number of merging galaxies at great cosmic distances and observations of the formation of galaxy clusters from smaller components.

One very important diagnostic tool for researching the structural evolution of the intergalactic medium, from which the galaxies are formed, is the spectrum of absorption line systems. Light from very remote sources such as quasars passes through a series of gas clouds on its way to us. The clouds can be located in the outer regions of galaxies, for example, or inside younger protogalaxies or the intergalactic medium between the galaxies. The gas absorbs certain components of light and leaves a characteristic signature behind in the spectrum of the quasar (Figure 2.6). The spectra that can be observed with the modern instrumentation on the VLT are of laboratory quality and allow detailed diagnostics of the gas clouds with regard to physical parameters such as density,

Absorption lines and the intergalactic medium

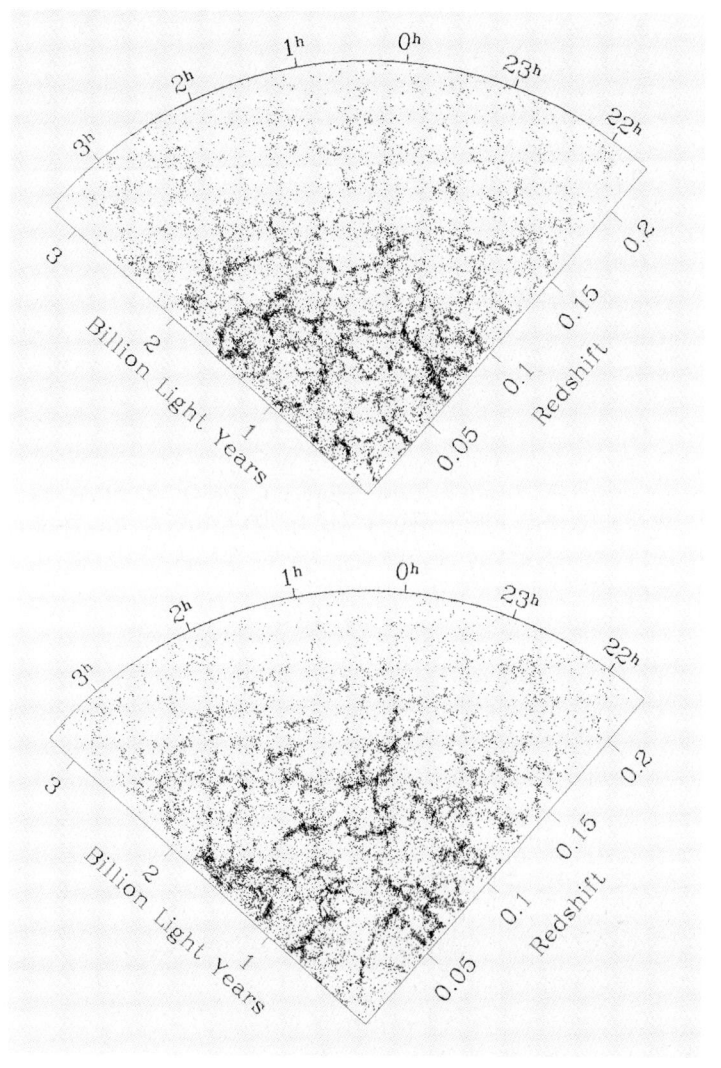

Fig. 2.5: Comparison of real galaxy distribution, as observed in Australia by the 2dF sky survey (top), with the simulated galaxy distribution in the Hubble volume simulation. This supercomputer analysis of the structural evolution of the universe, the world's largest to date, was carried out in Germany (bottom). (MPA)

temperature, turbulence and chemical composition. Because the clouds are located at various distances from us, we see them at different stages in the evolution of the universe. Quasars with a redshift of around 6 have been spectroscopically investigated using the VLT; they emitted their light at a time when the universe was only 5% of its present age!

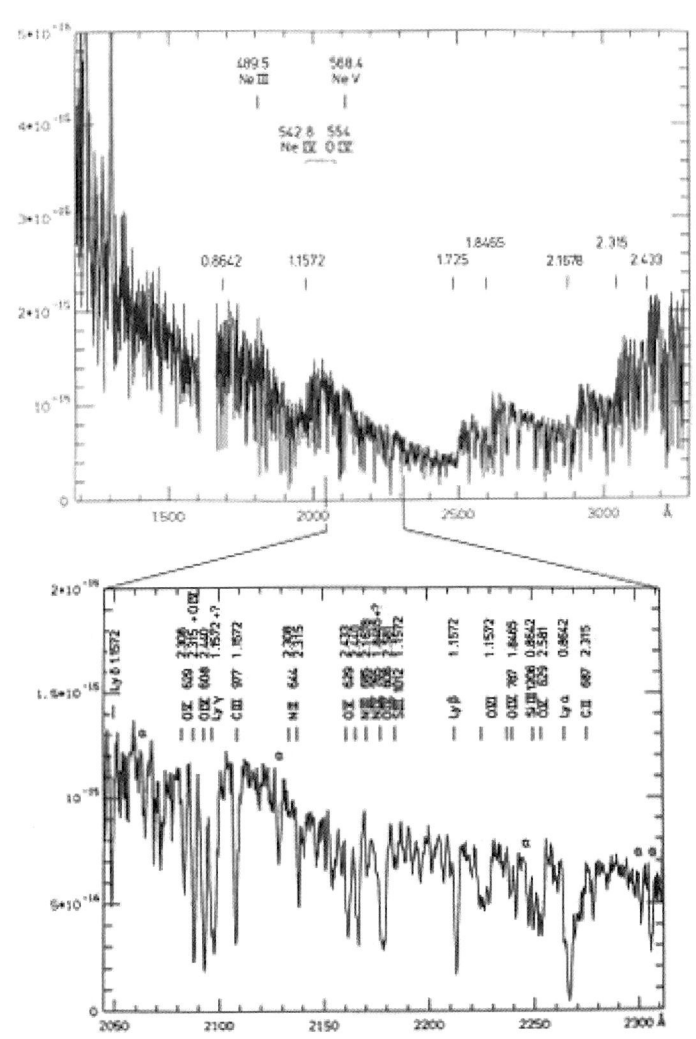

2.1 The universe – its origin, evolution and large-scale structure

Fig. 2.6: Tomography of the intergalactic medium along the line of sight to the quasar HS 1700+6416. The strength of the numerous absorption lines in this UV spectrum captured by the HST (the horizontal axis represents the wavelength in Ångströms) provides detailed information on the element abundance and the physical conditions in the thinly distributed gas in the foreground of the quasar. This gas still displays a very "old" chemical composition, which has not yet been widely enriched by the nucleosynthesis taking place in massive stars. (Source: NASA/HST/Hamburg University)

2 The scientific issues

The Lyman Alpha Forest

Whilst the development of dark matter can today be well traced numerically, the corresponding development of normal, baryonic matter is considerably more complex, because it is subject to additional interactions. On the other hand, the distribution of baryonic matter is accessible to direct observations. In recent years some preliminary advances in the numerical simulation of these components of matter have been made, leading, among other things, to the solution to an old mystery: deciphering the nature of the "Lyman-Alpha series". By this we mean a large class of absorption lines that can be observed in every quasar with a large redshift. This phenomenon, which has been known for 20 years, can today be theoretically followed using computer simulations by passing light rays through the simulated gas distribution of the early universe and taking hydrogen absorption into consideration. These synthetic spectra cannot be statistically differentiated from those of observed quasars. The Lyman-Alpha series therefore reflects the distribution and physical state of the gas in the early universe.

Primordial gas

Most recently, observations of the Lyman-Alpha series grew in importance when it became possible to measure the ratio of hydrogen to deuterium in the young universe. Even now it is assumed that we have penetrated far enough back to observe the primordial frequency ratio, as originally generated by nucleosynthesis in the big bang. This, in turn, is decisively governed by the cosmic density of baryonic matter. It is thus possible to determine this density by deciphering the chemical composition: it amounts to only around 4% of the critical density. In conjunction with the results of investigations of the microwave background, it can be seen that beside this normal type of matter, there are other components of matter that dominate the density of the universe to a far greater degree.

Such observations were only made possible by telescopes of the magnitude of the VLT. The large-scale survey of the sky by the Hamburg observatory is of crucial importance for selection of the requisite brightest quasars. More recently, the Sloan Digital Sky Survey, in which German researchers are participating, also provides a constant stream of new, very remote quasars. Future observations will make it possible to study the development of the chemical and hydrodynamic structure of the absorbing gases in great detail and thus to also gather important data on the early phase of the formation of galaxies and large-scale structures.

2.1 The universe – its origin, evolution and large-scale structure

A decrease in the density of the absorption line systems as the redshift decreases can be observed in the spectra of quasars and, parallel to this, in the formation of new galaxies. This corresponds to the concept that new galaxies are formed from the absorbing gas. Simultaneously, identification of the frequency of heavy elements in the absorption line clouds allows predictions to be made concerning the chemical enrichment of the intergalactic gas. This provides important insights into the rate of stellar genesis and the enrichment of the intergalactic medium with heavy elements. In particular, traces of heavy elements can also be found in absorption line systems at very large redshifts, indicating a very early epoch of stellar formation. Direct observations of this first stellar generation represent one of the most interesting challenges of observational cosmology and one of the primary objectives of the JWST.

Evolution of heavy elements

2.1.4 The large-scale structure of the universe

It is regarded as a fundamental truth that galaxies are not uniformly distributed around the universe. Rather, they collect together in the densest regions to form groups, galaxy clusters and superclusters. They have elongated or flat structures, which in turn enclose holes – regions of very low galaxy density – giving the distribution of galaxies in the universe a "bubble-like" structure when viewed on a scale of up to a billion light years.

The distribution of galaxies

For two decades now, much effort has been expended on surveying the structure of the universe, and these efforts are currently peaking in a study of the distribution of galaxies being carried out by the *Sloan Digital Sky Survey* (SDSS). These surveys confirm that galaxies, and probably all matter, are well structured on scales of up to several hundred million light years. For example, the discovery of the "Great Wall", a supercluster more than 500 million light years across, caused considerable commotion. Another example is the "Great Attractor", also a supercluster, which was originally noticed because of its gravitational effects on other galaxies. At least part of this supercluster has now been identified in optical images (Figure 2.7).

One of the original objectives of numerical simulations of the formation of cosmic structures was to compare the ensuing matter distribution with the observed distribution of galaxies. Theoretical considerations lead to the expectation that the distribution of galaxies would generally follow the distribution of

Comparisons to simulations

2 The scientific issues

Fig. 2.7: This image of part of the "Great Attractor" supercluster was made using the WF1 wide field camera on the ESO/MPG-2.2-m telescope. (ESO/MPG)

dark matter, tracing it. This has been verified by investigations of the dynamics of galaxy movements. In fact, the numerically generated distribution of dark matter is remarkably similar to that of the galaxies (see Figure 2.8). This, in turn, can only be regarded as an impressive confirmation of the prevalent cosmological model. In coming years, the SDSS will allow the relationship between dark matter and the distribution of galaxies to be more precisely measured.

Analysis of the spatial distribution of galaxies and their movements allows conclusions to be drawn on the mean density of the matter in the universe. The large surveys currently being carried out have shown, on the one hand, that the density of dark matter equals about 30% of the critical density; on the other hand, the density of baryonic matter is only around one tenth. If this result is combined with the measurements of the anisotropies in the microwave background, which imply that the total density is equal to that of the critical density, it appears to confirm that the presence of vacuum energy as the third component of the cosmic substrate agrees very well with

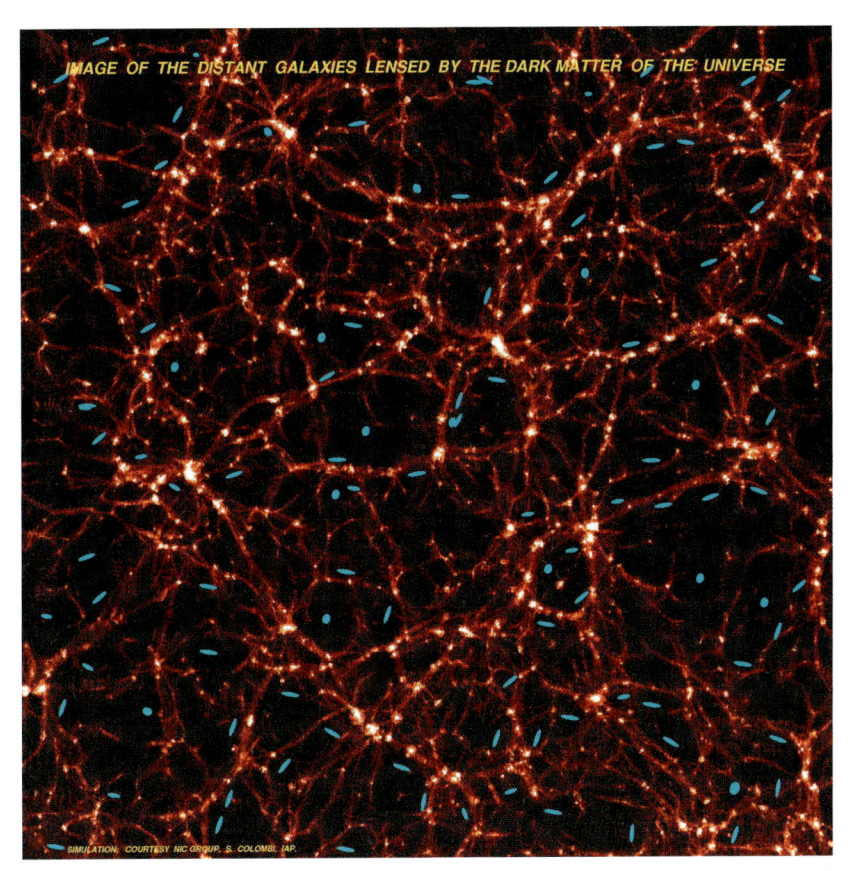

2.1 The universe – its origin, evolution and large-scale structure

Fig. 2.8: Supercomputer simulations of the origin of galaxies, galaxy clusters and large-scale structures. The picture shows the structure in the vicinity of the Milky Way. The variously coloured points show the positions and colours of the stellar populations of model galaxies as well as their positions relative to the underlying dark matter, shown in grey. (MPA)

the results of the Ia supernova observations. Furthermore, the baryonic density determined here correlates excellently with the value obtained from analyses of the chemical composition of hydrogen clouds, especially in conjunction with the theory describing the origin of elements in the early universe.

The observable images of far distant galaxies are distorted due to the deflection of light in the gravitational fields of cosmic structures. Because this so-called weak gravitational lens effect is determined by the total matter distribution, it is particularly suited to direct investigations of the distribution of dark matter. The effect is observed when investigating the shape and orientation of very faint galaxies. They populate the sky

Cosmic shear

so densely (typically, around 25,000 galaxies can be found in a region corresponding to the area of the full moon) that even very small distortions, which lead to a systematic alignment of the images of galaxies, can be statistically measured.

This effect, known as cosmic shear, models the density distribution of matter (Figure 2.9). At the beginning of the year 2000 four independent groups simultaneously published the initial results of their investigations, among them a German-French collaboration. The observed intensity of the distortions correlates wonderfully with the forecasts obtained from the cosmological standard model. The fact that these results notably confirm the models of the formation of structures in the universe is regarded as particularly important.

The data required for such investigations is acquired with the help of wide angle cameras. The enormous leaps made in the size of cameras in recent years have made such investigations possible. The OmegaCAM on ESO's new VST, and later the VISTA, provide first class research instruments to German astronomers, whose next target will be the determination of the cosmological parameters using methods completely independent of those used previously. Estimates of the individual distances to the faint galaxies are especially interesting; the method available, the so-called photometric redshift, has only become available with the advent of near infrared photometry. The data expected from satellite missions such as PRIME, for example, are practically without competition in this field.

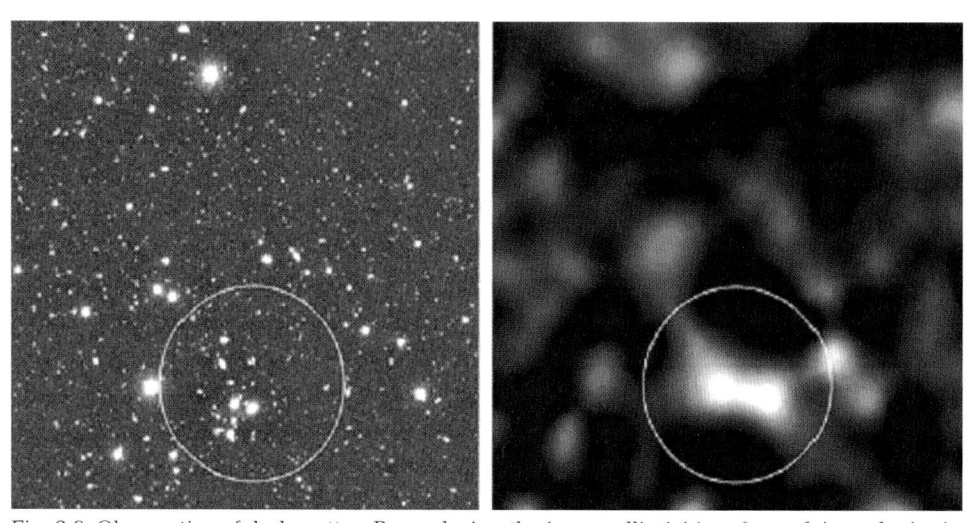

Fig. 2.9: Observation of dark matter: By analysing the image ellipticities of very faint galaxies in a deep VLT image (left), it was possible to reconstruct the mass distribution in this field (right). The position of the resulting mass concentration (circled) corresponds to a cluster of bright galaxies in the image. (ESO/Bonn University)

Box 2.1: The German X-ray satellite ROSAT

One of the great successes of German astronomy during the last fifteen years is the X-ray space telescope ROSAT. This satellite was launched in Cape Canaveral in 1990 by DARA/DLR under the overall leadership of the Max Planck Institute for Extraterrestrial Physics. It was developed in cooperation with Zeiss and Dornier and employs a Delta II rocket. ROSAT carried out the first complete survey of the sky using an imaging X-ray telescope and discovered more than 100,000 individual sources. Between 1990 and 1999, this highly successful mission made more than 4,000 individual measurements and generated almost 5,000 scientific publications. This makes ROSAT almost as successful, in terms of data yield, as the Hubble space telescope. The experience gained with ROSAT in the field of mirror and detector development were very successfully integrated in ESA's cornerstone mission XMM-Newton and also represent the foundation for the planned development of XEUS and ROSITA.

2.1.5 Galaxy clusters as cosmic laboratories

Galaxy clusters are the most massive gravitationally collapsed systems in the universe. Their evolution timescale of more than a billion years is not much smaller than that of the universe itself; clusters are therefore dynamic, young systems, occasionally revealing their evolutionary history. They form a bridge, so to speak, between cosmology and "classical" astrophysics and are intensively investigated using a wide range of methods and wavelength ranges. German institutes are involved in the leadership of a number of these research projects, which have capitalized on the capabilities of ROSAT and the VLT.

Galaxy clusters were originally characterised as condensation of galaxies; with the development of X-ray astronomy, it became increasingly clear that the space between the galaxies of each respective cluster is filled with hot gas at temperatures up to 100 million degrees and that the mass in hot gas exceeds that of the galaxies in the cluster. The dominant mass component of galaxy clusters, however, is dark matter, as Fritz Zwicky concluded in 1933 from his observations of the movements of galaxies.

The mass of galaxy clusters

This conclusion has now been unquestionably confirmed. Both the total mass and the mass profile of clusters can be determined by one of three entirely different methods. First, the observed movements of cluster galaxies allow their masses to be estimated, for which these movements represent a state of structural equilibrium. A second method consists of investigating the hot X-ray gas: it would evaporate like boiling water in a pan if it was not restrained by a very strong gravitational field. In fact, the X-ray luminosity and temperature of these gases correlate very closely with the total mass of the cluster (Figure 2.10). The third method employs the weak gravitational lens effect; the observable distortion of the images of background galaxies is greater, the larger the mass of the galaxy cluster.

Applying all three methods to a multitude of clusters in the past few years has revealed that only around 3% of the mass of clusters is found in the stars located in their galaxies and a further 17% in the hot gases between these galaxies, but that around 80% of the mass of galaxy clusters is located in dark matter.

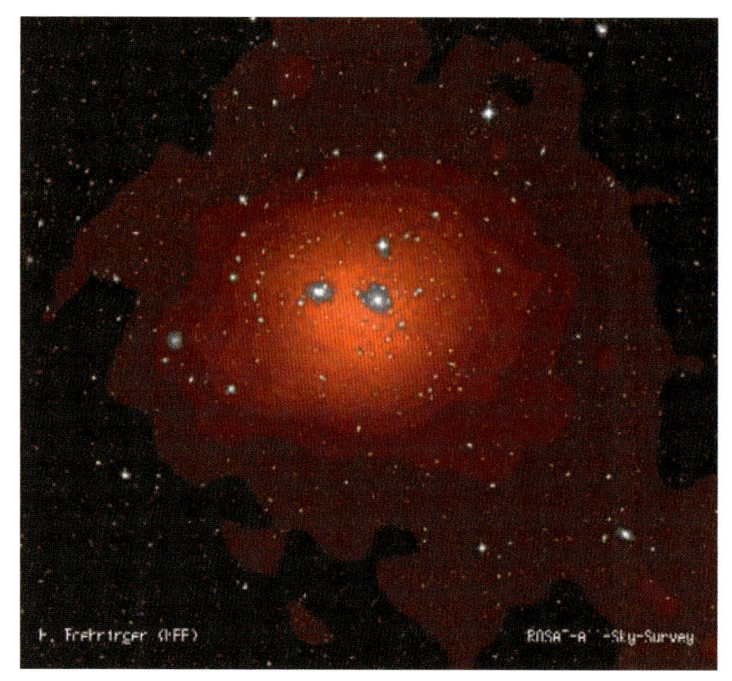

Fig. 2.10: A comparison of the optical and the ROSAT X-ray pictures of the Coma galaxy cluster. This optical picture taken from the Palomar Sky Survey is shown in grey scale and the observed X-ray emissions on a red background. (MPE)

2.1 The universe – its origin, evolution and large-scale structure

The evolution of galaxy clusters

All three methods of determining mass may also be applied to galaxy clusters with large redshifts, although multi-object spectrographs on large telescopes, powerful X-ray satellites and optical wide angle cameras are required here. It can be seen that galaxy clusters formed relatively early. This result, which was initially unexpected, can only really be understood if the density of dark matter in the universe is considerably lower than the critical density, as is the case in the currently preferred model of a universe dominated by vacuum energy.

The intergalactic medium

One major mystery involves the high frequency of heavy elements present in the intergalactic medium within the galaxy clusters. A substantially greater mass of gas and heavy elements can be found here than in all the galaxies of the respective cluster. The production of such large quantities of heavy elements cannot be understood by studying the evolution of our own galaxy, for example. Identifying the relative frequencies of heavy elements in remote galaxy clusters will contribute greatly to clarifying the origins of the elements. XEUS will open up new opportunities here, thanks to the X-ray spectroscopy of clusters in the very early universe.

Furthermore, the two new X-ray observatories Chandra and XMM-Newton have shown that highly complex interactions take place in the intergalactic gas of the clusters. In particular, it was possible to demonstrate the influence of the radio jet of a central active galaxy on the X-ray gas in some clusters by comparing X-ray images and detailed radio maps. Simultaneously, cooling gas in the cluster can serve as a fuel for the central engine – probably a black hole. These observations are also of enormous interest to researchers of magnetic fields in the intergalactic gas of clusters, and their interactions with the magnetised plasma of the radio jets. They also represent new challenges, especially to the field of hydrodynamic simulations.

Galaxy clusters and the large-scale structure of the universe

Similar to the distribution of galaxies, the distribution of galaxy clusters traces the large-scale structure of matter in the universe. Based on the ROSAT sky survey, it was possible to demonstrate this method for surveying cosmic structures. Even the relatively modest sample of 450 galaxy clusters displays clear density fluctuation structures at scales up to around two billion light years (Figure 2.11). One important objective for the future is therefore to extend these investigations to cover substantially greater volumes of the universe.

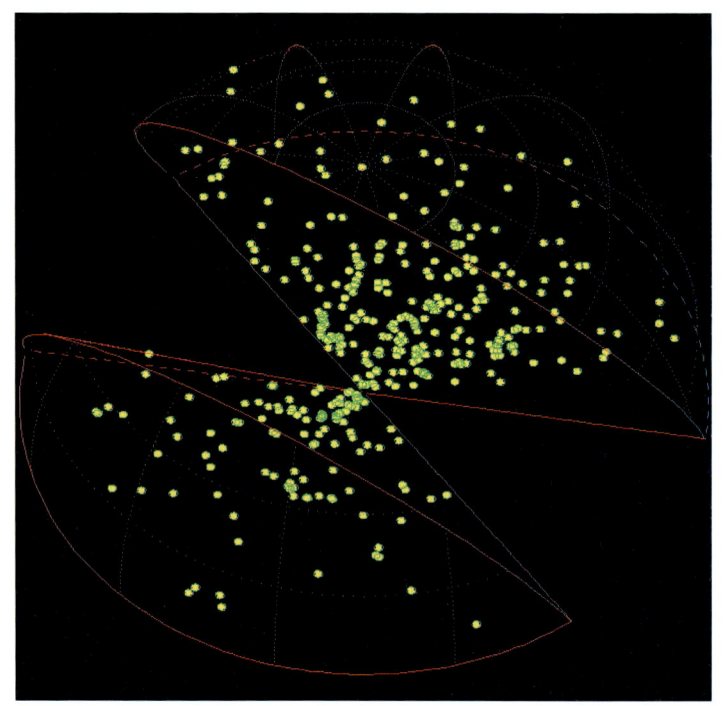

Fig. 2.11: Spatial distribution of a complete sample of the galaxy clusters discovered by ROSAT in the southern hemisphere. The obvious distribution structuring is clearly visible. The areas outside of the cone-shaped volume of space are hidden by dust in the disk of the Milky Way. (MPE)

The search for galaxy clusters

Efficient methods for discovering galaxy clusters are required in order to investigate their spatial distribution. Beside the traditional method of identifying excessive densities of galaxies in certain parts of the sky, a number of other search criteria are now available. The most prominent of these to date is the search for and identification of X-ray sources, as very successfully carried out using ROSAT data; ROSAT has so far identified more than 1,500 galaxy clusters. A similar survey of the XMM-Newton archive in the future will considerably extend this sample to the less luminous and more remote clusters. A sky survey with greater sensitivity than is available with ROSAT, and which would reveal a substantially greater number of clusters, has already been proposed as a national project and is being prepared under the name of ROSITA.

The weak gravitational lens effect offers a second, completely different and complementary method for finding galaxy clusters, because the gravitational lens effect is independent of the type of matter involved, in contrast to observations

2.1 The universe – its origin, evolution and large-scale structure

in the optical or X-ray range. The deflection of light around a concentration of mass generates a characteristic distortion pattern on the background galaxies, which are generally arranged tangentially to the mass centre. A systematic search for these circular "shear fields" allows a cluster population to be defined based solely on their gravitational properties. A number of galaxy clusters have already been located in this way; consistent application of this method using the new VST will uncover a large number of clusters. A comparison with galaxy clusters found using optical or X-ray observations will allow conclusions to be reached on the range of properties displayed by galaxy clusters. For example, one interesting question is whether the relative proportion of normal (baryonic) matter to the total matter within the cluster population varies greatly. This would be an indication of extreme events in the evolution of large-scale structures.

The third new method of searching for galaxy clusters is based on the so-called Sunyaev-Zel'dovich effect. It occurs when the cosmic background radiation passes through the hot gas in the interior of a galaxy cluster before it reaches Earth. This makes itself known by a typical signature in the radio and submillimetre range of the background radiation. This effect will initially be observed with the help of the *Atacama Pathfinder Experiment* (APEX) in particular, a German-led ALMA pilot project. Thousands of galaxy clusters are also expected to be found in this way using the Planck space telescope. The Sunyaev-Zel'dovich effect also allows an alternative determination of the Hubble constant, independent of any distance scale.

2.1.6 Ultra-high-energy gamma-ray astronomy

In the future cosmology will increasingly be able to access observation data relating to areas that were previously only accessible with difficulty, if at all. These include gamma rays, which, due to their penetrating nature, can even be received from source regions that otherwise remain hidden when observing in other wavelength ranges, due to the strong absorption by gas and dust clouds.

New methods of observation

Using Cherenkov telescopes at the HEGRA installation on La Palma, an intense and variable flux of gamma quanta were observed originating in the active galaxies Mrk 421 and 501. The spectrum of the radiation from Mrk 501 displays exponential behaviour, cutting off at around 6 TeV (Figure 2.12).

Very high-energy gamma radiation from active galaxies

2 The scientific issues

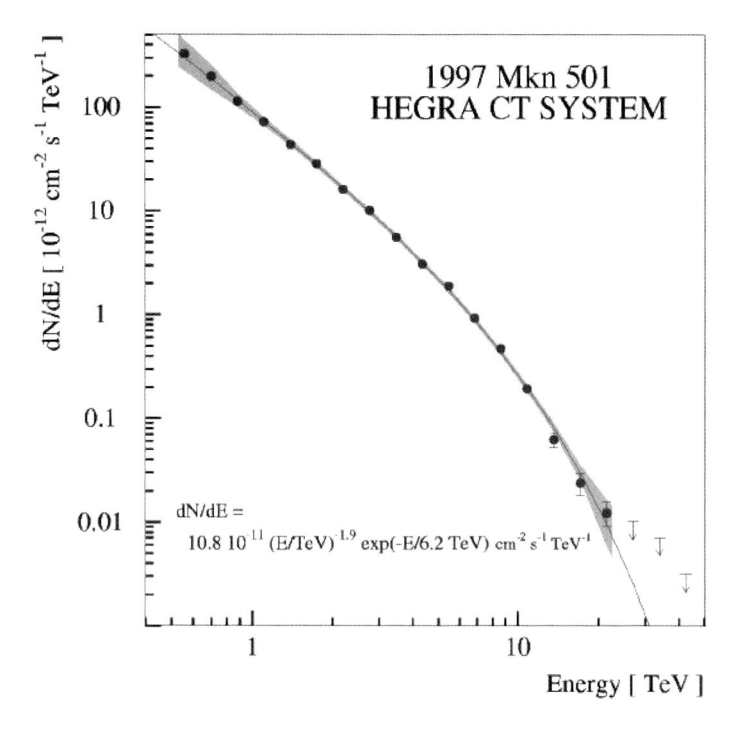

Fig. 2.12: The spectrum from Mrk 501 in the very high-gamma-energy range, measured using the Cherenkov telescope at the HEGRA installation. (MPIK)

[Figure: 1997 Mkn 501 HEGRA CT SYSTEM spectrum, dN/dE [10^{-12} cm^{-2} s^{-1} TeV^{-1}] vs Energy [TeV], with fit $dN/dE = 10.8 \cdot 10^{-11} (E/\text{TeV})^{-1.9} \exp(-E/6.2\,\text{TeV})$ cm^{-2} s^{-1} TeV^{-1}]

We will deal with the origins of this radiation, which is extremely interesting to observe from a cosmological perspective, in the next chapter. At the high energies involved, the gamma-ray photons (gamma rays) interact with the intergalactic radiation field in the infrared, optical and ultraviolet wavelengths, and the gamma quanta are converted to electron-positron pairs. Observations of the Mrk 501 source at energies in excess of 10 TeV therefore imply an upper bound to the energy density of the intergalactic radiation field. This allows an estimate to be made of how much energy galaxies and black holes have generated and radiated since the universe began.

Much more sensitive investigations of this type can be carried out with the next generation of instruments. They include the space observatory GLAST in the GeV range and the observatories H.E.S.S., MAGIC and other ground-based instruments with ranges from several 10 GeV up to several TeV. They will allow additional sources of very high-energy gamma-rays to be identified. Investigations of the origin of this radiation represent one of the focal points of gamma as-

2.1.7 Astroparticle physics

The properties of today's universe and its large-scale structures are directly linked to the properties of elementary particles. Within a fraction of a second after the big bang the latter determined the fundamental properties of the expanding universe, the nature of the inflationary phase and the initial conditions of structure formation. At this early time the interaction energy of the elementary particle fields was considerably greater than can be generated in Earth-bound accelerators. This explains the synergy of particle physics and astrophysics, because many of the most significant problems of very high energy particle physics can only be studied "experimentally" within an astroparticle physics framework. In fact, to date, all data on physics beyond the standard particle physics model has come from astrophysics.

Synergy of particle and astrophysics

In the recent history of the universe, too, high particle accelerations occur in some astrophysical objects, together with extremely high densities, which can be utilised to test particle physics hypotheses. The enormous astronomical distances involved allow the smallest differences in signal travelling times, as required by some theories of quantum gravitation, to be measured. The complementary perspective of studying the physics of astronomic sources on the basis of the particles emitted by them, is also highly relevant to astroparticle physics.

Cosmic particle accelerator

A series of experimental astroparticle physics techniques have been rapidly developed in recent years and open new windows to astrophysics: low-energy cosmic neutrino spectroscopy, precision measurements of gamma spectra in the TeV range and experiments involving searching for the elementary particles of dark matter have only recently become possible.

All known types of elementary particle were generated during the big bang. After a certain time expansion made further reactions unlikely, depending on the initial particle densities and mean interactions, and the corresponding species of particle was removed from the reaction cycle. Within a short time, the photons and neutrinos of the background radiation had been produced, as well as baryonic matter. However, it is also

Relics of the big bang

possible that various exotic, weakly interacting elementary (massive) particles (WIMPs) survived. As constituents of dark matter they may then have decisively influenced the further development of the universe, especially the formation of structures.

Experiments to detect WIMPs

Astroparticle physics uses a number of methods in its search for WIMPs. One of the detection processes used is based on the fact that WIMPs can be elastically scattered by atomic nuclei when they enter a crystal, for example. The WIMP then transfers a very small amount of recoil energy to the nucleus. Completely new and highly sensitive detectors needed to be developed and built in order to measure this energy. Moreover, they had to be screened against cosmic rays and other forms of interference. Detectors of this type are currently employed in the Gran Sasso underground laboratory in Italy.

The small scatter cross-section of the WIMPs requires crystal detectors weighing many kilogrammes, operated over periods of several years. For example, the DAMA experiment in Gran Sasso is based on a sodium iodide crystal weighing around 100 kg. The DAMA group reported a signal that may have been a indication of WIMPs. Larger versions of this experiment are under discussion for the coming decade. In Europe, these include the CRESST and EDELWEIS experiments, which use cooled detectors, and in which Germany is crucially involved, and the GENIUS experiment employing germanium counters. Together they cover a significant portion of the parameter space predicted by elementary particle physics (Figure 2.13).

WIMP accumulation in gravitational fields

A further technique for the detection of massive WIMPs is based on the accumulation of such particles in a gravitational field, for example in the centre of the Earth, the sun or in the centre of the Milky Way. Depending on the properties of the respective particles, the WIMP density in these regions may be enriched enough that they are subject to mutual annihilation. In the galactic core, this process of annihilation, thereby forming two photons, can lead to a flux of gamma quanta. They may be searched for using the next generation of Cherenkov telescopes (H.E.S.S. and MAGIC). The requirement is that the WIMPs must be located in a mass range between 100 GeV and several TeV. If WIMPs were annihilated in the centre of the Earth, the radiation produced would be detectable by the planned neutrino telescope ICECUBE.

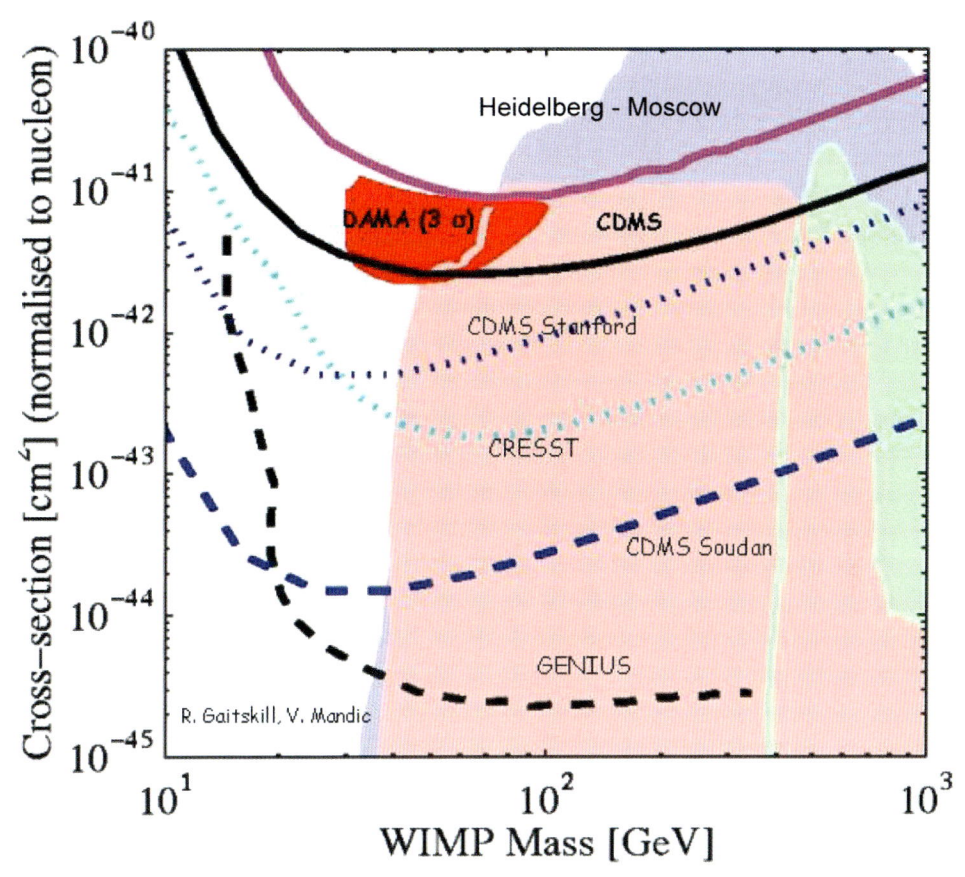

Fig. 2.13: Sensitivity of existing and planned experiments to search for dark matter. The exclusion zones are shown as a function of the assumed scatter cross-section and the WIMP mass. The points illustrate the WIMP parameters resulting from various model classes, the hatched region is the zone of potential detection by the DAMA experiment. (MPIK)

Topological defects

Another possible type of highly interesting relic of the big bang are the topological defects such as cosmic strings or magnetic monopoles. It is possible that these topological defects decay to form elementary particles with masses of a magnitude of 10^{24} eV. The particles themselves are also unstable and decay to form quarks, gluons and leptons. These topological defects may therefore represent sources of the very-high-energy particles of cosmic rays with energies of 10^{20} eV and more. The flux and the direction of origin of these hypothetical topological defect decay products can be investigated by the Pierre-Auger project or EUSO.

2 The scientific issues

2.1.8 New properties of neutrinos

Perhaps the most interesting particle physics results in recent years relate to neutrinos. These results are not produced by accelerator laboratories, but from astroparticle physics experiments.

Solar neutrinos

The principal components of solar neutrinos were first detected in the 1990s. On the one hand, the particle streams detected by the Homestake, GALLEX and SAGE experiments provide information on nuclear fusion processes occurring in the central regions of the sun (Figure 2.14). On the other hand, however, precise analyses have allowed far-ranging conclusion on the nature of neutrinos themselves. For example, it is now assumed that neutrinos of various families can mutually convert; they are said to oscillate. The observations made by the Japanese detector Super-Kamiokande point to the same conclusion.

These results signify a revolution in elementary particle physics and have important consequences for the choice of possible cosmological models, because neutrinos can only oscillate if they possess a rest mass. The previous standard particle physics model assumes neutrinos are massless.

Neutrinos and dark matter

However, the experiments carried out so far do not allow masses to be precisely determined. This question, nonetheless, is eminently important both for particle physics and for cosmol-

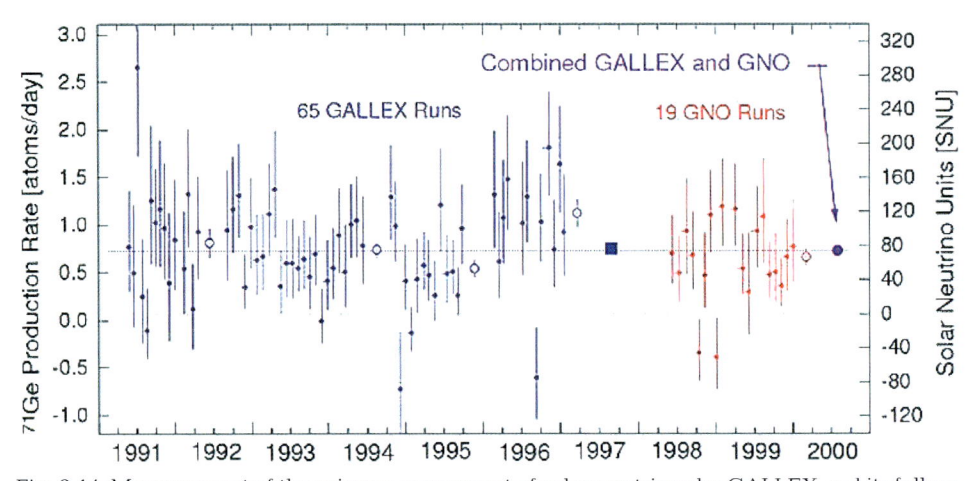

Fig. 2.14: Measurement of the primary component of solar neutrinos by GALLEX and its follow-up project GNO. The counting rates of the various measurement runs using these two neutrino detectors at the Gran Sasso laboratory are shown. Only around 2/3 of the expected neutrino flux of the sun is observed, clearly indicating a finite mass for the neutrinos. (MPK/TU Munich)

2.1 The universe – its origin, evolution and large-scale structure

Fig. 2.15: The counting test facility of the BOREXINO neutrino experiment. This solar neutrino detector is being built with German participation in the Gran Sasso laboratory. (LNGS)

ogy, because neutrinos were considered possible candidates for dark matter at quite an early stage. Currently, many physicists favour a class of models asserting that all neutrinos are too light to provide a significant contribution to dark matter and thus to the density of matter in the universe. Nevertheless, scenarios involving neutrinos that possess greater mass cannot be completely ruled out.

The mass of neutrinos and their associated magnetic moments are also relevant to a theoretical understanding of supernovae. Today, it is assumed that neutrinos actually facilitate the explosion of massive stars.

Future solar neutrino experiments

The objective of a new generation of experiments to detect solar neutrinos is therefore to determine the parameters for neutrino oscillations. In the high-energy field, this will be accomplished by the SNO (Canada/USA) and Super-Kamiokande (Japan) experiments. German working groups are participating intensively in the European high-energy experiments GNO, BOREXINO (see Figure 2.15) and the planned LENS experiment.

2.1.9 Gravitational wave astronomy

Detection of gravitational waves

Almost ninety years after the publication of Einstein's general theory of relativity, relativistic gravitational physics is changing from a purely theoretical field of research to an experimental science. In 1993 the American astrophysicists Joseph Taylor and Russell Hulse were awarded the Nobel Prize for Physics for their indirect proof of the existence of gravitational waves. They observed two pulsars in a mutual orbit for a number of years. The pulsars emit gravitational waves, thereby losing energy, and gradually approach one another. This in turn reduces their orbital period in exactly the manner predicted by the general theory of relativity. Direct detection of gravity waves by large laser interferometer gravitational wave detectors, several of which are approaching completion across the world, would not only spectacularly confirm Einstein's theory, and thus the foundation of modern cosmology, but also open a new observation window for gravitational wave astronomy.

Gravitational waves from the big bang

The detection of gravitational waves from the early universe, which is fascinating from a cosmological point of view, is a particularly difficult but fundamental aim of this young branch of research. As relics of an inflationary phase of the universe, these gravitational waves may have been generated similarly to the known density perturbations and therefore also contain fundamental information on the early phase of our universe. Today, these perturbations should form a gravitational wave background.

Some inflationary models predict typical spectra, which may be measured within the next five years by gravitational wave detectors on Earth. Other models suggest that the typical structures should appear in a frequency band only accessible to detectors in space. Estimates indicate that a single gravitational wave detector such as LISA should be capable of detecting a gravitational wave background at a frequency of 0.01 Hz. If the gravitational wave signal is weaker than the detector noise, it may be found by cross-correlating two independent detectors. At a frequency of 20 Hz, the LIGO II detector system in the American LIGO observatory, which is running roughly simultaneously with the Anglo-German GEO600 project, may be sufficiently sensitive.

2.2 Galaxies and massive black holes

Scientific advances

- Detection of massive black holes in the Milky Way and in the centres of nearby galaxies.
- Close relationship between the mass of black holes and that of galaxies.
- Detection of numerous young galaxies one to three billion years after the big bang.
- Clarification of the nature of the X-ray and infrared background.
- Cosmic evolution of stellar production and nuclear activity in galaxies.
- Origin of the Hubble sequence and its temporal evolution.
- Detection of dark matter in various galaxy types.
- Detection of high-energy gamma-rays from active galactic nuclei.

Purpose and aims

- To investigate the origins and evolution of galaxies in the early universe by direct observations.
- To investigate galactic evolution by drawing conclusions from observing today's galaxies.
- To discover the first generation of stars.
- To identify the role of black holes and active galactic nuclei in the structure and evolution of galaxies.
- To open new observation windows for gravitational waves, high-energy gamma-rays and neutrinos.

Galaxies as dynamic stellar systems

Galaxies are systems of stars, interstellar matter and dark matter, which are held together by gravitation (Figure 2.16). Gravity also acts between the galaxies and allows them to come together to form the largest known structures, the galaxy clusters and superclusters. These may extend across several hundred million light years (see Section 2.1). Our picture of galaxies has been dramatically altered and expanded during recent decades. Instead of relatively static and isolated "islands", we see them today as highly active and dynamic systems that can interact strongly with each other. For example, if they pass close to one another tidal forces act, which force stars and interstellar material from their paths. It is even possible for galaxies to merge. The structure and evolution of the galaxies represents an extremely interesting and active research field.

2 The scientific issues

Fig. 2.16: The spiral galaxy NGC 1232 is a typical spiral galaxy. It is similar to our Milky Way. (ESO)

Today, for the first time, enormous improvements in the limits of detection of faint objects allow stellar evolution to be studied on cosmic time scales and the effects of interaction processes that took place in early epochs to be directly observed. Using techniques available today, individual stars can be discerned in nearby galaxies, allowing the evolution of these objects to be directly compared to computer simulations covering periods of several billion years (Figure 2.17).

However, the beginning of this evolutionary process, the actual course of galaxy creation and production of the first generation of stars, is still widely unknown. These fundamental processes represent a central problem of astrophysics. The exact role of the various constituents of the physical process contributing to galaxy creation, and their interactions with one another, must be understood in order to investigate further. Only then can the evolution of the galaxies be deduced from their appearance today.

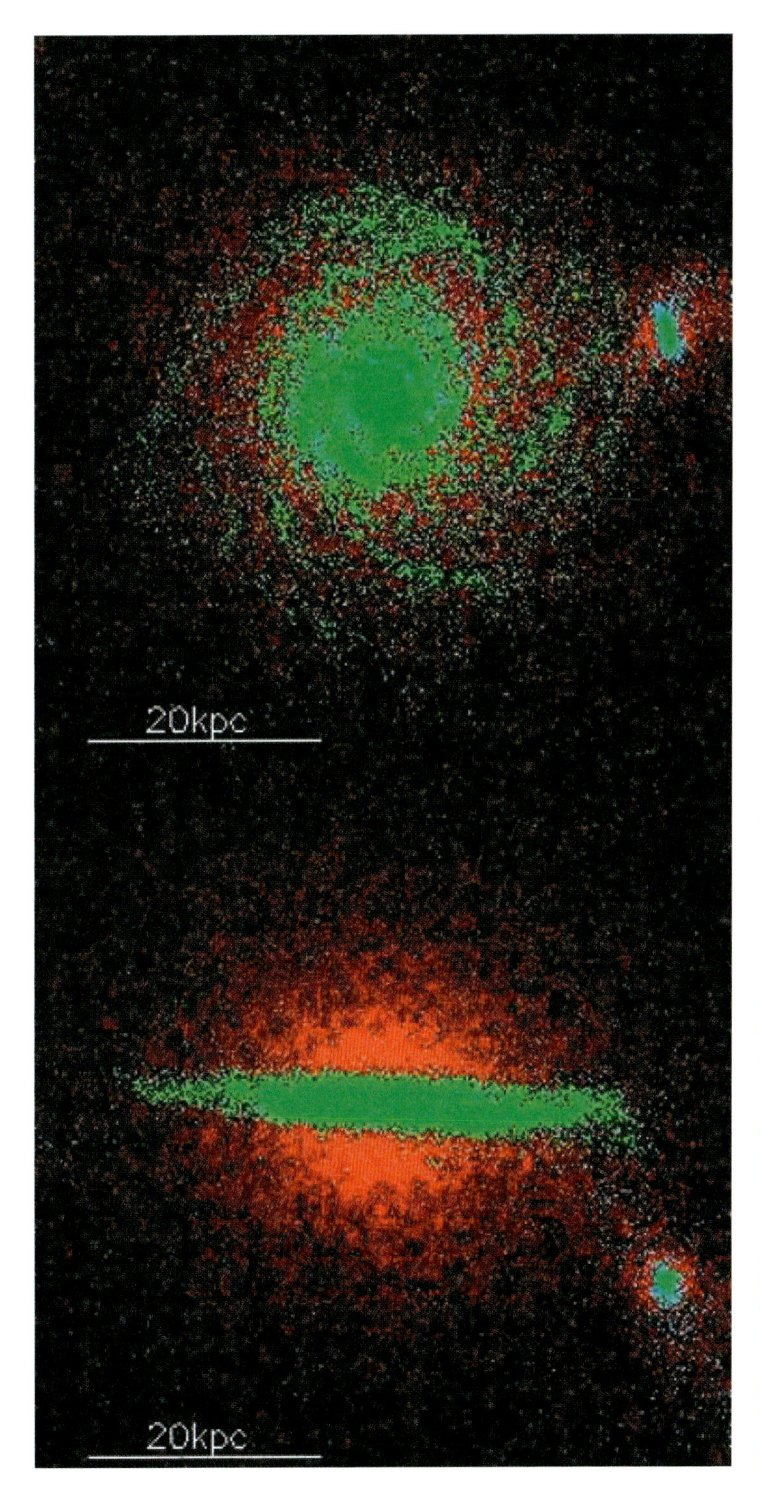

Fig. 2.17: Frontal and side view of a spiral galaxy, as it appears in the newest generation of high-resolution cosmological simulations. It shows the distribution of older stars (older than six billion years, red), young stars (younger than six billion years, blue) and gas (green). Whilst the older stars are spherically distributed due to the merging of the precursor galaxies, the young stars display a disk-like distribution similar to the gas from which they form. (AIP/Steward Observatory/University of Victoria)

2 The scientific issues

2.2.1 Origin and evolution of galaxies

Origin of galaxies

The models describing the formation of structures in the universe (see Section 2.1) are aimed not least at understanding the origin and evolution of galaxies. Whilst numerical simulations of dark matter in the universe are now quite advanced, and predict linked structures with the mass of galaxies, they cannot yet be observed. Only when the luminous portions of galaxies can be modelled will it be possible to directly compare theoretical forecasts with observations. The problem here, on the one hand, is that the processes to which gas is subject are considerably more complex than for dark matter. On the other hand, the length scales requiring investigation are so varied as to dismiss the hope of modelling the formation of galaxies from the cosmic substrate, and the formation and evolution of their stellar populations, by direct simulations. Despite this, we have now acquired a good qualitative understanding of the evolution of galaxies, thanks to the combination of numerical models of dark matter with semi-analytical models, developed in Germany, describing stellar populations (Figure 2.1). The models are based on highly simplified assumptions, but take the principal processes into consideration. Among others, these include gas falling into the halos of dark matter, the stellar production rate, the coupling of stars and gas by means of supernova explosions and enrichment of the gas with heavy elements. Naturally, these models are provided with a multitude of parameters, which must be specified on the basis of observations. By incorporating numerous new observations, it is expected that these models will be considerably refined in coming years, thus leading to a deeper understanding of cosmic evolution.

Dark matter in galaxies

We have known for a long time, from observing the rotation speed of stars and gas inside spiral galaxies, that spirals are surrounded by an extensive halo of dark matter. It dominates the dynamics of galaxies outside of the central zone. It has only been possible in the last few years to demonstrate conclusively that elliptical galaxies are also dominated by dark matter. Whilst the spatial extent of luminous matter can be observed, there are only few indicators of the extent of the dark halos. Because the total mass of a galaxy in particular is a function of the size of the halo, this is an especially important question. Considerable advances are expected here from investigating the gravitational lens effect on a large sample of galaxies. In the future, this will be possible with wide-angle surveys, for example using the VST.

2.2 Galaxies and massive black holes

Sensitive surveys with large area coverage

Investigating the evolution of galaxies requires significant samples of the universe at a variety of redshifts. A large proportion of our current knowledge is based on deep surveys in visible light, the near infrared and in the X-ray range. Spectroscopic investigations using the newest large telescopes and the *James Webb Space Telescope* (JWST) are necessary for a more exact understanding of the sources. The aims are to penetrate to even more distant stellar systems, to carry out detailed physical and dynamic investigations, and to observe the less luminous galaxies, these being particularly important in the hierarchical model.

The long-exposure and very deep-field images of the universe acquired by the Hubble space telescope (Figure 2.18), ESO's *Very Large Telescope* (VLT), the Anglo-American-German X-ray satellite ROSAT, the European infrared space telescope ISO and the new submillimetre array detectors SCUBA on Hawaii and MAMBO on the IRAM 30 m telescope, are especially relevant. These observations allow the contribution of the early galaxies to the extragalactic background radiation to be determined for a broad sample of the electromagnetic spectrum. However, in order to eliminate the influence of absorption, star formation can only be measured in the submillimetre and radio ranges. In particular, the new large telescopes allow the spectra of very faint galaxies to be measured in the visible and near infrared wavelengths and their distances (or, more precisely, their spectral redshifts) and physical properties to be determined.

Past investigations are afflicted by the fact that the observed sections of the sky were too small. It is important that future surveys cover wide areas, for example using the VLT Survey Telescope. This will provide representative samples containing a sufficient number of galaxies. Only such large samples will also contain the exotic "key objects". Even the local "census" of dwarf galaxies, which are regarded as relics of the original phase of galaxy formation, can be further complemented. Detailed study of these objects provides information complementary to direct observations of the early phase of galaxy formation.

Much seems to point to the fact that the first stars and galactic nuclei were produced at redshifts of $z > 6$. The birthplaces of stars and galaxies are probably extremely rare, not especially luminous and can only be found in the infrared due to their large redshift. Identification of these real protogalaxies and protoquasars demands searches covering hundreds of square degrees and down to very small flux limits of 1 to 3 µm. This is practically impossible from the ground but could be

Fig. 2.18: The deepest sky image of the universe to date was taken in the constellation of the Big Dipper (Ursa Major) using the Hubble space telescope. It shows a multitude of galaxies at various distances and stages of development. (NASA/STScI/ESA)

achieved using a space telescope such as PRIME, for example. The protogalaxy candidates found in this way can be subject to a detailed spectroscopic investigation in the near infrared using the JWST. X-ray spectroscopy using XEUS can also detect the earliest protoquasars.

The extragalactic background radiation

In the past, a radiation field extending over various wavelength ranges was detected in the sky; it is probably a relic of galaxy-forming activity in the early universe. It contains around twice as much energy in the infrared than in the visible light range

2.2 Galaxies and massive black holes

Box 2.2: Infrared, submillimetre and millimetre astronomy: a new field for German astronomy

Infrared, submillimetre and millimetre astronomy has evolved enormously since the beginning of the 1980s: the sensitivity of detectors, cameras and spectrometers has increased continuously. German groups have played a leading role in these developments. The German-French-Spanish institute IRAM (Institute for Radio Astronomy in the Millimetre Range) has provided German researchers with access to the world's best telescopes in the millimetre wavelength range for more than 10 years. During the 1990s the European Space Agency's (ESA) *Infrared Space Observatory* (ISO) formed a further highlight. Between 1995 and 1998, during its surprisingly long life of 29 months, the telescope, cooled to only a few degrees above absolute zero, made around 26,000 individual observations in excellent quality. It was thus the world's most successful infrared mission of the 1990s. German institutes (and industry) were crucially involved in building the detector and participate in exploiting the astronomical data and in data archiving. Substantial new observations and information on the interstellar medium, the origin of stars in our Milky Way and on stellar production activity in external galaxies, including objects in the early universe, was gathered using these instruments (Figure 2.19).

(Figure 2.20). The nature of this radiation has so far only been partially deciphered. In the low-energy X-ray range, ROSAT was able to resolve almost the entire background into individual objects (see Figure 2.21): these objects are almost entirely very remote quasars and active galaxies. XMM-Newton and Chandra are investigating the important energetic X-ray component and subjecting the sources to spectroscopic analysis. Only in this way can they be unequivocally identified, and their distances determined.

ISO in the infrared, and SCUBA und MAMBO in the (sub)millimetre wavelengths, have found the first remote, dust-rich galaxies with high stellar production rates. These measurements, as well as follow-up observations using the VLA and the IRAM mm-range interferometer, show that the important, energetic "submillimetre population" of remote galaxies is not visible in the optical range and is almost invis-

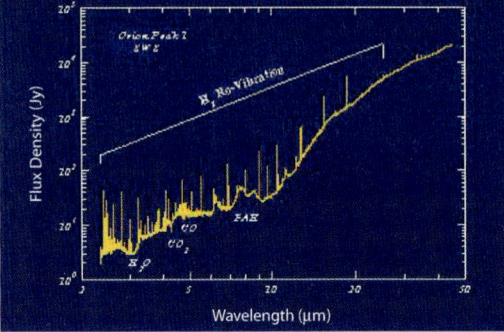

Fig. 2.19: Infrared, submillimetre and millimetre astronomy. Right: The ISO satellite in the ESTEC testing facility before launch in 1995. Top left: IRAM's Plateau de Bure interferometer, consisting of five 15 m telescopes, is currently the world's most sensitive interferometer facility in the millimetre range. Bottom left: One of the many results of the ISO mission were the first detailed, spectroscopic measurements in the infrared. Shown here is the mid-infrared spectrum of the Orion star formation region. Many lines of H_2, CO and other molecules emanate from the dense and warm gas, which is excited by the young stars embedded in it. These lines provide detailed information on the physical and chemical processes in regions of stellar production. (IRAM/ESA/MPE)

ible in the near-infrared. The submillimetre telescopes APEX, and later Herschel and ALMA, are therefore essential to further studies of the nature of these objects. Because of the infrared/radio correlation such galaxies can then be detected in the radio range with SKA. Similar difficulties (extremely weak emissions in the optical range) exist for many of the sources discovered by XMM-Newton and Chandra.

The deepest images demonstrate that in the early period of the universe catastrophic collisions and merging galaxies were considerably more common than they are now. Stars form here at a rate more than ten times that which we observe in the universe today. The massive black holes that were present in the compact galactic nuclei even then must have been fed a large amount of matter, so that around one hundred

2.2 Galaxies and massive black holes

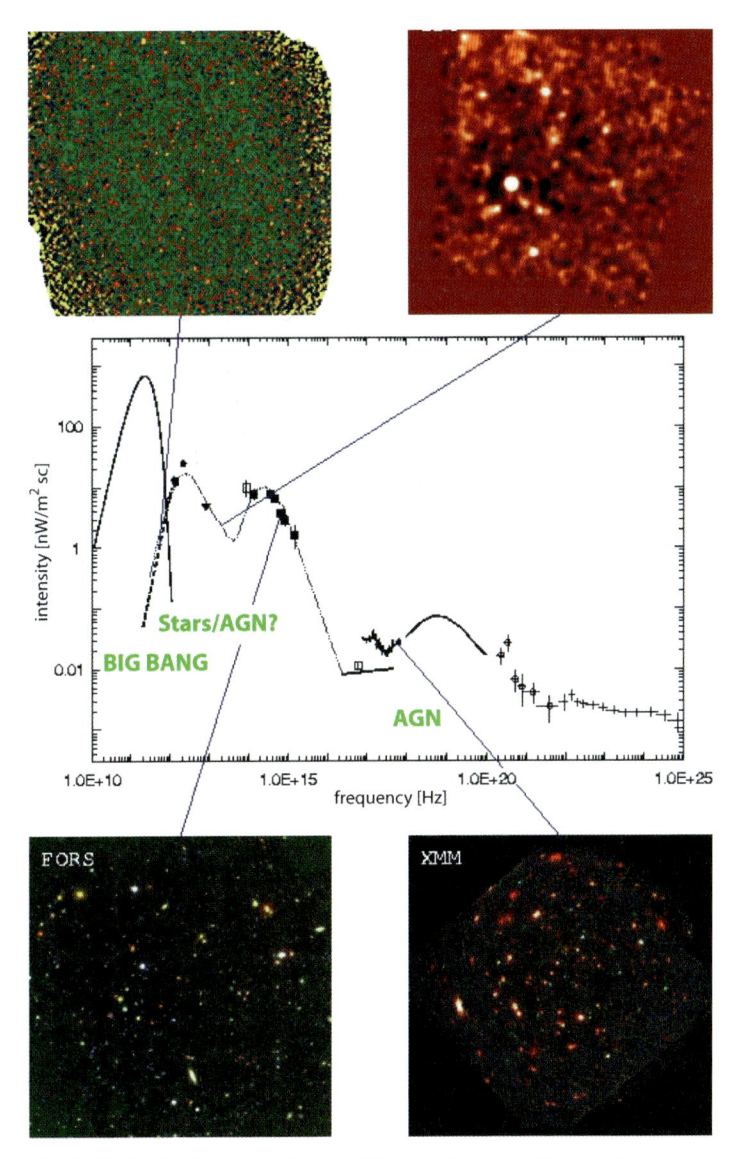

Fig. 2.20: In the last decade, satellites and ground-based observatories have begun to resolve the cosmic background over large areas. The deep surveys shown here, including German, ESO and ESA projects, range from millimetre frequencies (Mambo), through infrared (ISO) and visible light (FORS), into the X-ray range (XMM). (AIP/MPE/MPIfR/ESO/ESA)

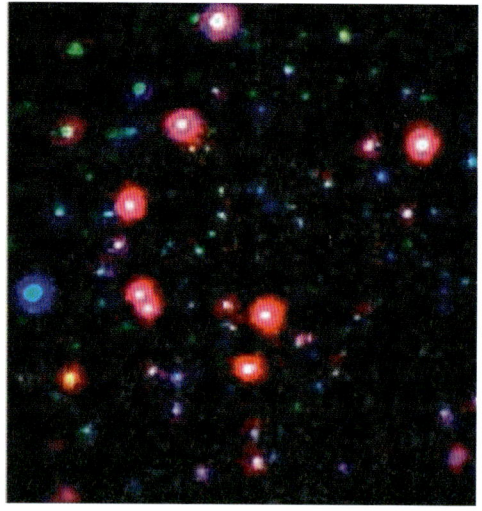

Fig. 2.21: Left: The ROSAT satellite flight model in the space testing facility. Right: One of the most important scientific results of the ROSAT mission was resolving the mysterious X-ray background radiation, which has been known for many years, into a finite number of single sources. Most of these weak sources turned out to be very remote quasars; the different colours represent the hardness of the X-ray spectrum. (MPE/AIP/DLR)

times more galaxies were active than today. The interplay of activity and stellar production and their relative weighting represent very important factors. It can also be observed in relatively close, luminous galaxies.

In the future, the *Atacama Pathfinder Experiment* (APEX) will bring further advances, but it will be the Herschel space telescope, with its extreme sensitivity, which will resolve the far infrared cosmic background into individual sources. Surveys carried out using the telescope will locate a large number of star-forming and active galaxies in the early stages of the universe, when most of the heavy elements were formed and the number of active galaxies was at its greatest. Considerable progress is expected on somewhat longer timescales, especially in the sensitivity and spatial resolution of ALMA in the submillimetre range, and in X-ray spectroscopy using XEUS. The close relationship between far-infrared radiation and the radio continuum also allows investigation of stellar production in remote galaxies without interference from absorption. However, the sensitivity and resolution of the *Square Kilometre Array* (SKA) are necessary for this (see Figure 2.22).

Nucleus activity and stellar production

It is only since recently that we really have a concept of the course of stellar production in the universe. This has been made possible by identifying the redshift (distance) of objects

2.2 Galaxies and massive black holes

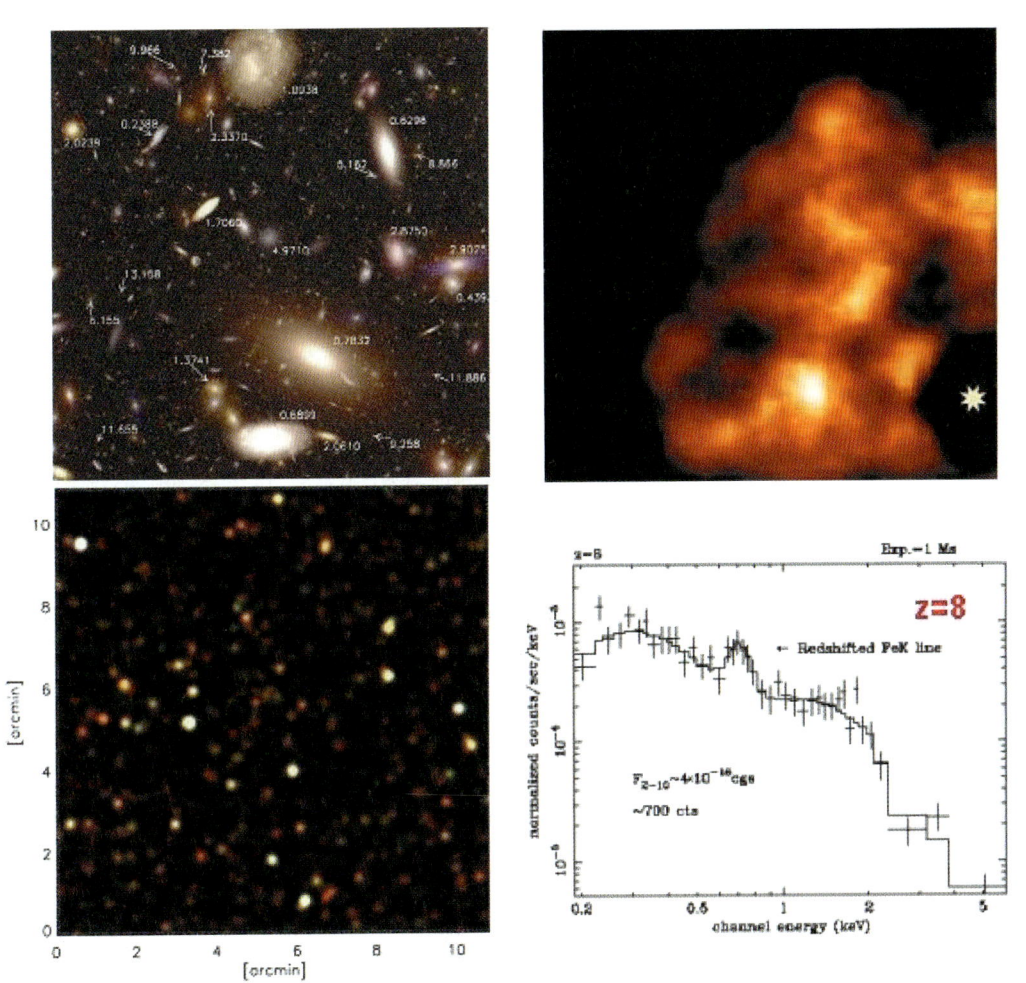

Fig. 2.22: Observations of sky objects in the young universe. The top left figure shows a simulation of a deep JWST image in the near infrared, where the numbers represent the redshifts of variously distant galaxies. This image penetrates one or two orders of magnitude deeper than the most sensitive measurements possible today. The JWST observations can see young galaxies, very bright globular clusters and supernovae at redshifts up to 5-10 if they emit ultraviolet and optical radiation in their frame of reference. In contrast to this, deep surveys by Herschel (bottom left: simulation of an approx. 110 square arc minutes field) will discover dusty galaxies and active galactic nuclei at z = 1-4 and resolve a large proportion of the far-infrared background into individual sources. Top right: Simulation of an SKA observation of an atomic hydrogen cloud in the intergalactic medium at a redshift z ~ 9 (about 600 million years after the big bang), which is being illuminated and excited for the first time by a newly formed quasar (star in the lower right corner) at the end of the cosmic Middle Ages. Bottom right: X-ray spectroscopy of the remotest black holes by XEUS. A simulation of the 0.3-20 keV spectrum of an active galactic nucleus at z = 8 is shown. XEUS can measure the characteristic emission lines, providing information on the masses of the first black holes and the element abundances in their surroundings. (Sources: NASA/JWST, SKA, MPE and MPIfR)

located previously during deep sky surveys. Through observations using ROSAT and from quasar surveys in the optical and radio ranges, we also know more now about the matter falling into active galactic nuclei.

These observations show that the stellar production rate and the activity in galactic nuclei developed in a surprisingly similar way, even though the level of activity in the early universe was more than ten times greater than it is today. This, too, is a further indicator for a close relationship between the origin of galaxies and the massive black holes at their centres. The deep X-ray and infrared surveys planned for the future will make it possible to retrace in detail the evolution and the links between stellar formation and active nuclei as far back as the early universe, and to better understand the possible causal relationships. More information on remote galaxies is also expected from gamma astronomy. The spectrum of a remote gamma source similar to those observable by H.E.S.S. and MAGIC for example, contains information on the total energy released by galaxies and black holes in space between a given gamma source and Earth.

Reconstruction of stellar production in today's galaxies

Beside direct observation of the history of stellar production in remote galaxies, an "archaeological" perspective is also possible: the history of stellar production can be reconstructed for relatively close galaxies with the aid of the high-resolution images provided by the Hubble space telescope and the VLT, by analysing the brightness and colour of the stellar populations. With the future availability of adaptive optics on large terrestrial telescopes, as well as the JWST, this will be possible for much larger volumes than today. Dwarf galaxies are the most important subjects of these studies.

The accumulation of heavy elements in gas and stars, representing a cumulative index of the history of stellar production in galaxies and galaxy clusters, can also be investigated to a much greater extent by optical and X-ray spectroscopy. The background is: it is known that only the light elements, primarily hydrogen and helium, were created during the big bang. But today we can observe the complete spectrum of chemical elements in stars and in the interstellar medium. These were only "spawned" later in the interiors of stars and then ejected into space, for example in supernova explosions (see next section). One indicator of the evolutionary development stage of a galaxy is therefore given by the accumulation of heavy elements in its interstellar gas and stars.

Closely linked to the chemical evolution of the galaxies is the very significant question of when the first stars formed. Because they are formed from the (primordial) gas created during the big bang, they cannot contain any heavy elements. To date, not one of these first generation stars, which, as far as we know today, must have formed before the universe was 1/10 of its current age, has been found. It is hoped to make a first discovery with the JWST, for example by observing supernova explosions in this first stellar generation. But it is also possible that the remains of the oldest stars still populate the outer regions of the Milky Way, called the halo, in the shape of white dwarfs. A new, sensitive survey of the galaxy can therefore also contribute to answering this important question.

The first stellar generation

2.2.2 Structure of galaxies

As long as 80 years ago, Edwin Hubble introduced a classification system that has remained valid in its core until today. Put simply, it differentiates elliptical galaxies, two types of spiral galaxies and irregular galaxies. Despite various efforts, it has not yet been possible to clarify unequivocally the reason for these morphological differences.

Great progress has nevertheless been possible in recent years in investigating the structural parameters of galaxies, where German researchers have made valuable contributions. For example, it was discovered that a relationship exists for elliptical galaxies between the extent of the central region, the luminosity of the central region, the distribution of stellar velocities and the frequency of heavy elements. The new large telescopes allow these parameters to be determined even for remote galaxies. The question of how the relationships found for nearby galaxies change systematically for more distant ones will provide essential understanding of the evolution of galaxies. In this context the *Sloan Digital Sky Survey* is crucial to German researchers, because it allows the properties of today's galaxy population to be determined with great precision.

In recent years polarising observations in the radio continuum have shown that almost all galaxies possess magnetic fields, and that they are generally arranged in spiral patterns and have strengths of 5 to 10 µGauß (Figure 2.23). However, their influence on stellar creation and the formation of spiral arms remains largely a mystery. The sensitivity and resolving power of the next generation of telescopes are required to answer this question.

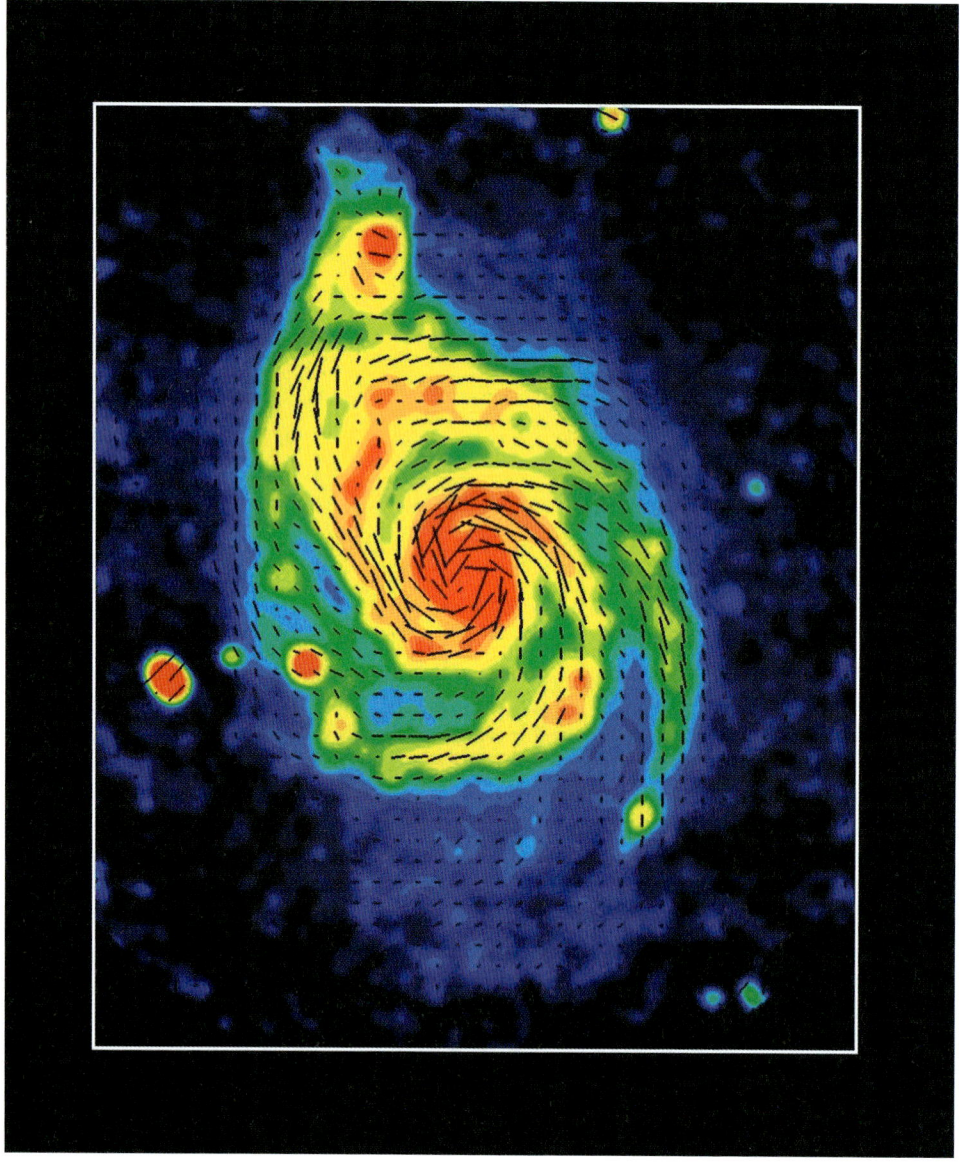

Fig. 2.23: Radio continuum image of the M51 galaxy with polarisation vectors at a wavelength of 6 cm. They show the orientation of the magnetic fields relative to the spiral arms. (VLA/Effelsberg)

The influence of bars in spiral galaxies

The structures occurring in disk galaxies – such as spirals and bars – can be regarded as the result of the intrinsic vibrations of a multi-component stellar system. The type of gravitation potential, density waves, the relationship between stars and the interstellar medium, and other processes, are subject to

2.2 Galaxies and massive black holes

Fig. 2.24: Galaxy NGC 1365, as seen here with the VLT, is a typical representative of a barred spiral galaxy. (ESO)

complex interactions. In recent years, by observing the Milky Way in particular, it has been possible to increase our understanding of the importance of bars (Figure 2.24) for the kinematics of the gas in the central regions of disk galaxies. The gas flux, which is influenced by the bar potential, leads to strongly non-radial velocity components, in turn leading to redistribution of the gas masses in the central region. The gas can promote stellar production and supply active nuclei with new matter. Polarised radio wavelength radiation in the centimetre range shows that the magnetic fields follow the gas flux in bar galaxies, providing us with new insights into the gas dynamics of galaxies. Numerical simulations of galaxies indicate that the distribution of stars may also be influenced by the bar potential.

Interactions between galaxies

Many of the properties of galaxies display a dependence on their environment. Morphology, for example, is a function of galaxy density: in regions with high galaxy densities the ratio of elliptical to spiral galaxies is greater than in regions of lower density. This can be explained by interaction processes

– either by merging or by gravitational interactions during close passes – with neighbouring galaxies (Figure 2.25). For example, observations in conjunction with numerical simulations have strengthened the hypothesis that elliptical galaxies are created by two large spiral galaxies merging (Figure 2.26). However, it remains unclear whether all elliptical galaxies are really produced in this way.

The deep exposures disclosed by the Hubble space telescope for example, (Figure 2.18) reveal that gravitational and merging interactions were the dominating elements of galactic evolution in the early universe. The most remote known galaxies today display only minor similarities to the galaxies in our region, and often the interactions of various galaxies can be directly recognised. Theoretical understanding of these processes signifies just the beginning and represents an immense challenge to numerical simulations. Because the numerical density of galaxies in clusters is substantially greater than the mean density, clusters are best suited to studying the interactions and evolutionary processes involved. It has been known for a number of years that the galaxy population in clusters

Fig. 2.25: The interacting galaxy pair NGC 6872 and IC 4970. (ESO)

2.2 Galaxies and massive black holes

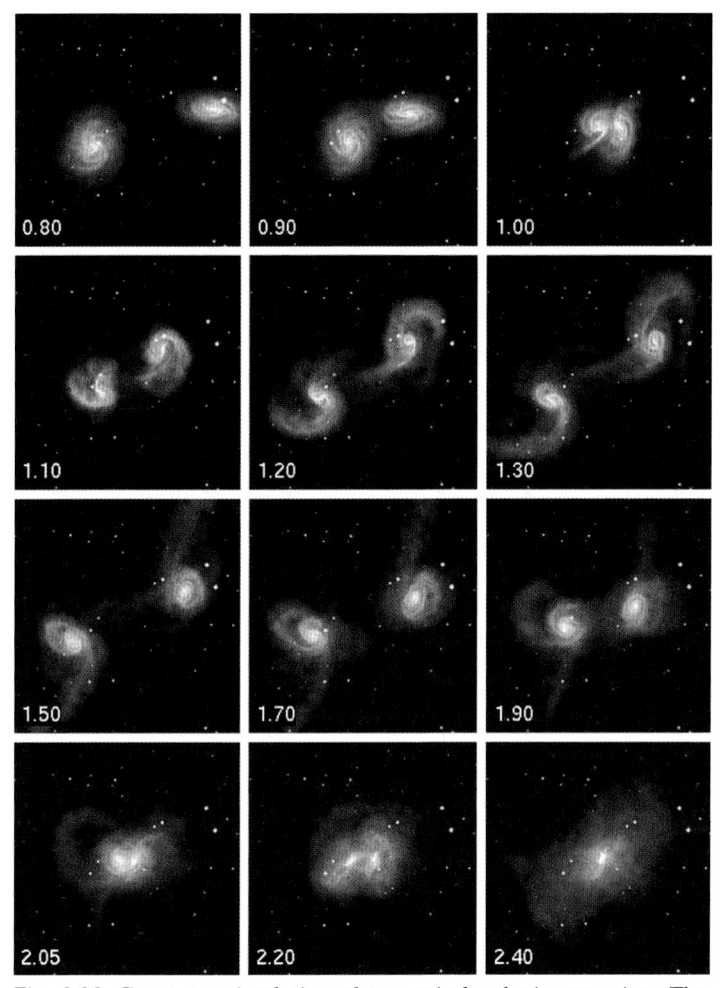

Fig. 2.26: Computer simulation of two spiral galaxies merging. The numbers indicate the time elapsed since beginning the simulation in billions of years. Once both systems have united an almost spherical stellar system remains: possibly an elliptical galaxy. (MPA)

changes with cosmic time, although little is yet known about the governing processes.

Theoretical models

The genesis and evolution of galaxies are closely related to the dynamics of the many-body system of stars and the various gas components influenced by gravitational forces. The recoilless many-body system and the complex dissipative physics of interstellar matter give rise to new phenomena, previously unknown to classical gas physics. The dynamics of such a recoilless many-body system are considerably more complex than gas dynamics for instance, leading to new phenomena, previously unknown to classical gas physics. This is one of the reasons why it is not easy to draw conclusions on the mass of an enclosed black hole from the dynamics of the stars in a galactic nucleus. Tests of the theories and models of the evolution of galactic stellar dynamics are therefore required. On the one hand the "laboratories" used are numerical simulations and, on the other, real objects such as clusters and galactic disks. Interactions of small-scale plasma processes with the large-scale dynamics of gas and stellar components by way of exchanges of mass and energy represent an important aspect of future research.

It would be almost impossible to understand complex systems such as the clusters surrounding black holes in galactic nuclei without the methods tested on these objects. Computer simulations allow systems to be prepared to allow processes and hypotheses to be tested, similar to a real experiment. Intense use of powerful computers, for example at federal computing centres, is essential to this purpose. Just as important is that German theoretical astrophysics maintains its hold in the field of specially adapted computers.

Some important problems facing our understanding of galaxies in the coming decade comprise the problem areas of stellar and gas dynamics, the structure of the Milky Way and galactic disks, the amount of dark matter, and the dynamics of dense galaxy centres and galactic interactions.

2.2.3 Massive black holes

We now know that the centres of many nearby galaxies contain concentrations of dark matter. In all probability these are black holes with masses of millions or even billions of solar masses. This awareness, to which German researchers have contributed extensively, came unexpectedly. The genesis of these black holes, and their link to galactic genesis, are not yet understood. Black holes have previously only been detected in two types of galaxies: in elliptical galaxies and in spiral galaxies possessing a stellar core region (bulge). This core is an almost spherical region containing predominantly older stars (Figure 2.27). Its shape and colour resemble those of an elliptical galaxy. No black holes have yet been detected in pure disk-shaped spiral galaxies, without a central bulge.

Surprising correlations

An astounding and interesting correlation between bulges and elliptical galaxies was recently discovered: the mass of a

2.2 Galaxies and massive black holes

Fig. 2.27: The disk of spiral galaxy NGC 4565 shows itself to us exactly edge-on. The spherical bulge can be recognised in the central region. (ESO)

black hole increases in a roughly linear manner with the absolute luminosity of the total stars in the bulge. If we assume that the luminosity is a measure for the amount of matter present as stars, the mass of the black hole thus increases with that of the nucleus, and the black hole always comprises around 0.2% of the mass of the surrounding nucleus. Whilst this correlation still displays a relatively large degree of dispersion (Figure 2.28, left), a much stricter relationship was found between the mass of the black hole and the velocity dispersion of the stars in the nucleus (Figure 2.28, right): the more massive the black hole, the faster the stars in the mother galaxy revolve. These

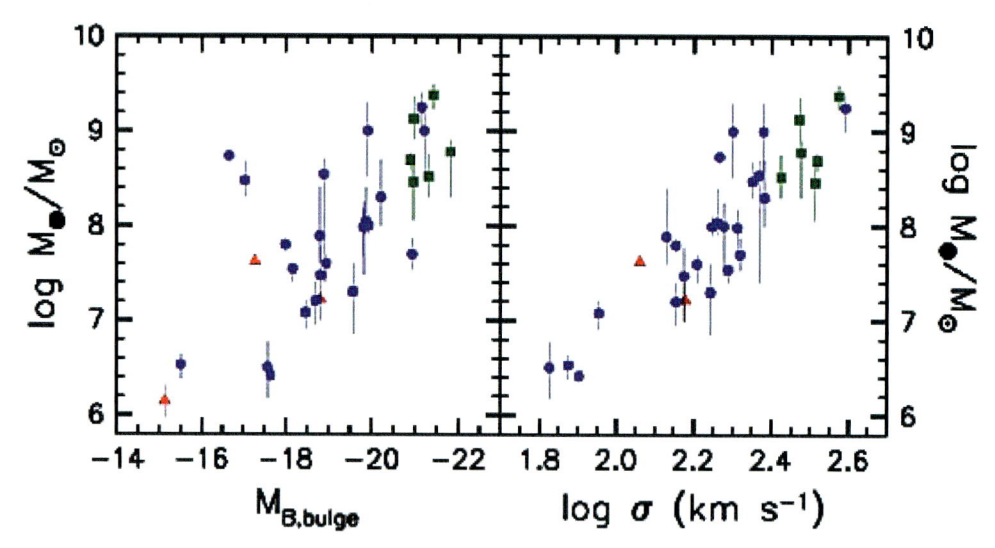

Fig. 2.28: Correlation of the derived masses of central black holes in nearby galaxies (vertical axis) with the mass (or luminosity, left) and the velocity dispersion (right) of the mother galaxies. This correlation indicates that black holes and galaxies formed together in a common process in the early universe. (NASA/HST/Munich University)

relationships are very unusual, because as the black holes are considerably less massive than the bulges, their gravity is also correspondingly smaller than that of all the stars in the nucleus. As a result, the black holes are not even capable of commanding the movements of the majority of stars in the nucleus. The majority of stars do not "feel" the effect of the black hole's gravitational pull.

We currently have only vague ideas of the reasons for these relationships. However, according to the models developed by German theoreticians, they may be an indication that black holes and galaxies evolved together. In the future, high-resolution observations using adaptive optics on the new large telescopes should not only allow unequivocal confirmation of nearby black holes, but should also better demonstrate the feedback processes between matter influx to black holes on the one hand, and stellar genesis and the role of galaxy merging on the other.

The black hole in the centre of the Milky Way

German astronomers are at the international forefront of studying the black hole at the centre of our Milky Way. Using Speckle interferometry and adaptive optics, they were able for the first time to three-dimensionally survey the movements of individual stars in the centre of the Milky Way to within a

2.2 Galaxies and massive black holes

few light days, and to model their radio emissions to scales of less than 15 Schwarzschild radii. The observations point to the existence of a compact central object of three million solar masses. This can only be a black hole (Figure 2.29).

An additional convincing demonstration of the existence of a massive black hole was possible with the aid of observations of the spatial distribution of the maser lines of molecular water vapour in galaxy NGC 4258 by intercontinental radio interferometry. Here, it was possible for the first time to detect a compact accretion disk rotating around a central point mass. NGC 4258 and the galactic centre are therefore currently the two best elements of proof that the dark mass concentrations in many nearby galaxies actually represent black holes.

It should be possible within the next decade to derive the orbits of the innermost known stars in the galactic core and thus to measure the gravitational field three to five times

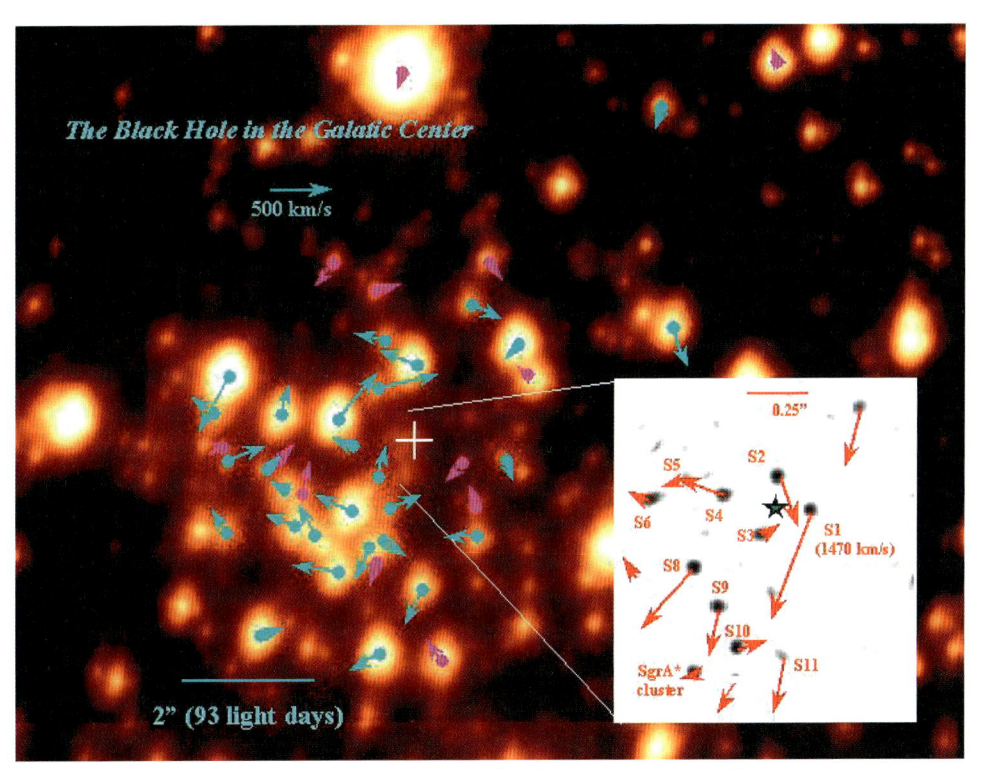

Fig. 2.29: The black hole in the centre of the Milky Way. The arrows indicate the measured intrinsic movement of the stars in the central region; the central 40 light days around the compact radio source SgrA* (green star) can be seen in the lower right box. The gravitation potential, which is dominated by a central point mass of ~ 3 million solar masses at the position of SgrA* can be derived from the stellar velocities. (MPE)

closer to the centre than previously possible. Investigations using the interferometers being built at the VLT and the *Large Binocular Telescope* (LBT) should achieve resolutions of only a few thousandths of an arc second and be sensitive enough to discover considerably fainter stars closer to the centre compared to today. It would be especially interesting to penetrate to the region of around 0.01 arc seconds from the centre, because it is here that the deflection of starlight caused by gravitational lensing should be noticeable. If this could be observed it would make it possible to determine the mass of the black hole with much greater precision. Using *Very Long Baseline Interferometry* (VLBI) in the radio wavelengths, it can already be shown today that radio and submillimetre radiation emanates from a very compact region only marginally larger than the assumed extent of the black hole itself. In the near future it should even be possible to directly observe the "shadow of the black hole" in the galactic centre. Theoretical estimates and the most recent X-ray observations show that the X-ray radiation must also originate in the direct neighbourhood of the black hole's event horizon.

X-ray observations of black holes

In recent years it has been possible for the first time to detect unusual phenomena in the gas disks of black holes using X-ray spectroscopy, in particular with the Japanese space telescope SCA. One iron emission line was observed, for example, that occurs in the laboratory at 6.4 keV. In the active galaxy MCG 6-30-15 it displayed a heavily expanded and asymmetrical shape (Figure 2.30). This can be explained by effects of both the special and the general theory of relativity that only occur in the immediate vicinity of a black hole. Currently, very detailed observations of these iron lines are being carried out on many nearby active galactic nuclei (AGN) using the European X-ray telescope XMM-Newton. The details of the X-ray spectra have proven to be far more complex than predicted by previous theoretical models. Precise investigation of the temporal variability of the line character using XMM-Newton, and later the X-ray observatory XEUS, will provide direct evidence of the mass and angular momentum of the black hole, as well as of the mass flux rate into the black hole. This would not only allow the innermost workings of the actual "motor" of an AGN to be studied for the first time, but would simultaneously provide an opportunity to test the general theory of relativity as it relates to very strong gravitational fields.

Another important diagnostic tool for the dynamics of the relativistic matter flux in the immediate vicinity of the black hole is given by investigation of the rapid temporal vari-

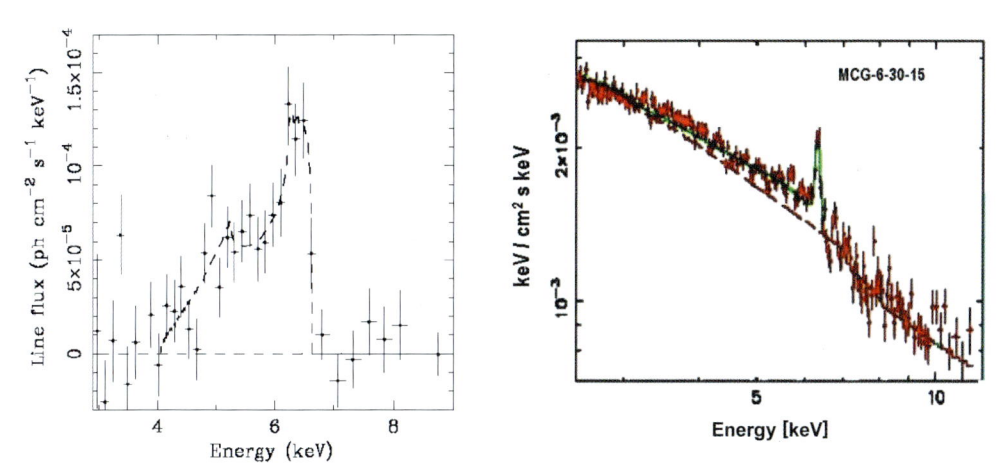

Fig. 2.30: The relativistically broadened X-ray line of iron in the active galaxy CG- 6-30-15. Left, as measured by ASCA, right by XMM-Newton. (ISAS/Tübingen University/ESA)

ability of X-rays and the more energetic gamma-ray radiation (several TeV). Simultaneous observations across a wide range of wavelengths and optimal use of the high-resolution radio interferometer are required in order to study the dynamics and instabilities of the gas flux.

Merging black holes

As mentioned above, the collision and merging of galaxies represent cosmic "traffic accidents" that have played a role in the evolution of many galaxies. But this poses the question of what happens to the black holes located at the centres of these galaxies. If the galaxies merge in such a way that their central black holes both remain in the new core area, the dynamic friction with stars will eventually bring them close enough together and lead to merging.

According to the theory of relativity, gravitational waves should be radiated during this process (Figure 2.31). The planned Euro-American gravitational wave detector in space, LISA, will be sensitive enough to register two massive black holes merging anywhere in the universe. LISA will probably be capable of detecting and resolving the details of several such events each year. The characteristic vibration frequency of gravitational waves is so low that such events can only be observed from space. Good samples of the expected signals are required from supercomputer simulations in order to interpret the actual detected signals. These complex simulations are the subject of intense efforts in Germany.

But cosmic objects, too, can be employed for direct and indirect gravitational wave measurements. Pulsars, rapidly ro-

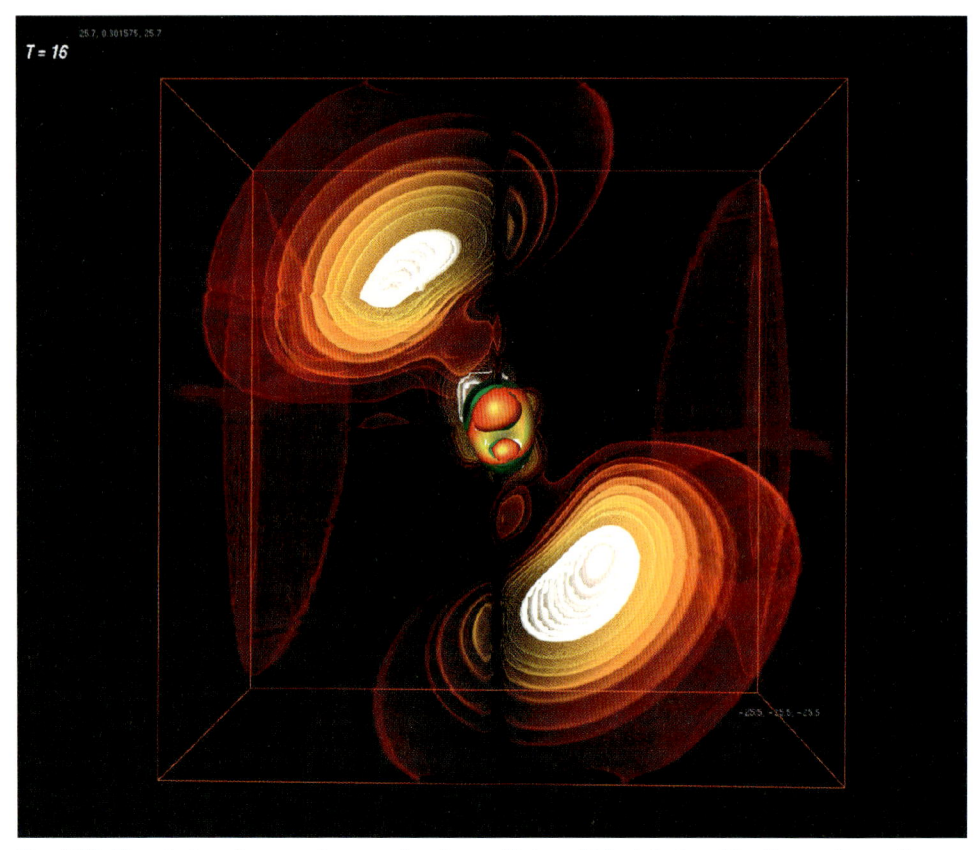

Fig. 2.31: Gravitational waves from a glancing collision of black holes. The figure shows the numerical simulation (the equations of the general theory of relativity on a supercomputer) of the gravitational wave emitted during the collision of two black holes. The black holes are represented by their "apparent horizons", roughly corresponding to the event horizons. The colours on the apparent horizons characterise the curvature of the surfaces: green represents a strongly positive curvature, red a spherical curvature (corresponding to a spherical surface), yellow is flat, cyan a negative (hyperbolic) curvature. The delicate envelope-like structures represent various stages of intensity of the emitted gravitational wave. Such collisions will be directly detectable using the space-based gravitational wave interferometer LISA. (NSCA)

tating neutron stars, are the most precise clocks in the cosmos; with their aid it is possible to measure the curvature of space. By precisely determining the time of arrival of the pulses and their variations, long gravitational waves, such as those emitted by close binary stars, can be detected. The SKA will play a decisive role in this endeavour in the future.

Possible effects on theoretical physics

The expected proof of gravitational waves is of fundamental importance to physics, over and above the mere astronomical applications. For the first time, these observations will make the details of the predictions made by the general theory of relativity testable, as they relate to the effects of very strong gravitational fields. Investigations of gravitational waves emitted when smaller black holes fall into larger ones will allow precise conclusions to be drawn about the gravitational field near the larger black hole. These tests of strong gravitational fields are of crucial importance, especially to theoretical physics. They may possibly also influence the experiments attempting to unify gravitation and the other interacting forces (strong and weak force and the electromagnetic interaction).

2.2.4 Active galactic nuclei

The explanation of active galactic nuclei (AGN) as accreting massive black holes can today be regarded as an extremely successful paradigm. It is applicable to a wide range of masses and luminosities and even covers considerably weaker forms of activity. In every third galaxy in the nearby universe we find visible signs of "mild" activity in their central regions. Extreme forms of core activity, which occur in a few percent of known galaxies, emit up to a thousand times more radiation than a "normal" galaxy. In addition, this radiation originates in an extremely small volume with the diameter of a single solar system. These are the quasars. So-called "unified models" are also generally accepted; they interpret the many different manifestations of the AGN phenomena as the consequence of the varying perspective of the observer to a structured and partially hidden galactic nucleus region. The dense gas and dust toroid around the active nucleus predicted by these "unified" models has now been confirmed by optical polarising measurements and radio observations of a number of objects. However, many aspects still pose questions.

For example, the relationship between the absorbing structures at various wavelengths (for example in the X-ray and UV/optical ranges) remains unclear. The exact spatial structure of the "toroid" requires observations with substantially greater spatial resolution, in particular with adaptive optics on large telescopes. Beside the accretion disk, relativistic plasma beams, also called "jets", have an important role to play in today's standard model. They shoot out of the disk, driven by magnetic fields, and withdraw mass and energy from the accretion disk. Moreover, it is possible that jets directly draw rotational energy from the black hole. The jet emissions extend across an extremely wide frequency range, from radio through optical and X-ray to the gamma and TeV range. High-energy particles (electrons, positrons, hadrons) play an important role (Figure 2.32). The structure of jets and their relativistic plasma

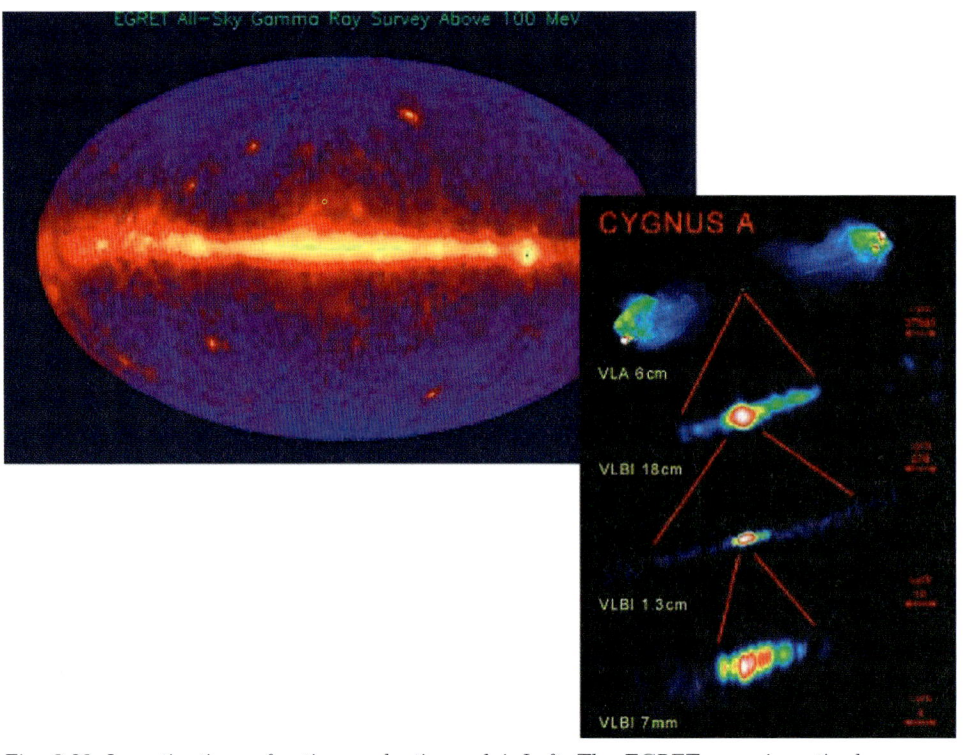

Fig. 2.32: Investigations of actives galactic nuclei. Left: The EGRET experiment's sky survey using the Compton Gamma Ray Observatory discovered a number of compact, extragalactic gamma sources, beside intensive, hard gamma radiation from within our own Milky Way's interstellar medium (horizontal strips). This gamma radiations originates in the relativistic jets of active galactic nuclei (NASA/DLR/MPE). Right: High-resolution measurements of the complex structure of a jet covering almost 5 orders of magnitude in a three-dimensional scale in the galaxy Cygnus A using intercontinental radio interferometry. (MPIfR)

can even be observed down to subparsec scales with the help of VLBI techniques (Figure 2.33). The central regions around black holes will in future be more closely investigated thanks to advances in submillimetre interferometry and moves towards larger baselines by way of Space VLBI.

With the adaptive optics on the new large telescopes and the millimetre range interferometry possible with the planned ALMA observatory, new opportunities will open up for highest-resolution investigations of the black hole's gas supply, which is currently only partially understood (Figures 2.34, 2.35).

Active galactic nuclei and stellar genesis

Another phenomenon, which has been hinted at by observations with the Hubble space telescope and in the near-infrared, will also be investigated. In some cases, areas of enhanced stellar genesis have been observed in the environs of active

2.2 Galaxies and massive black holes

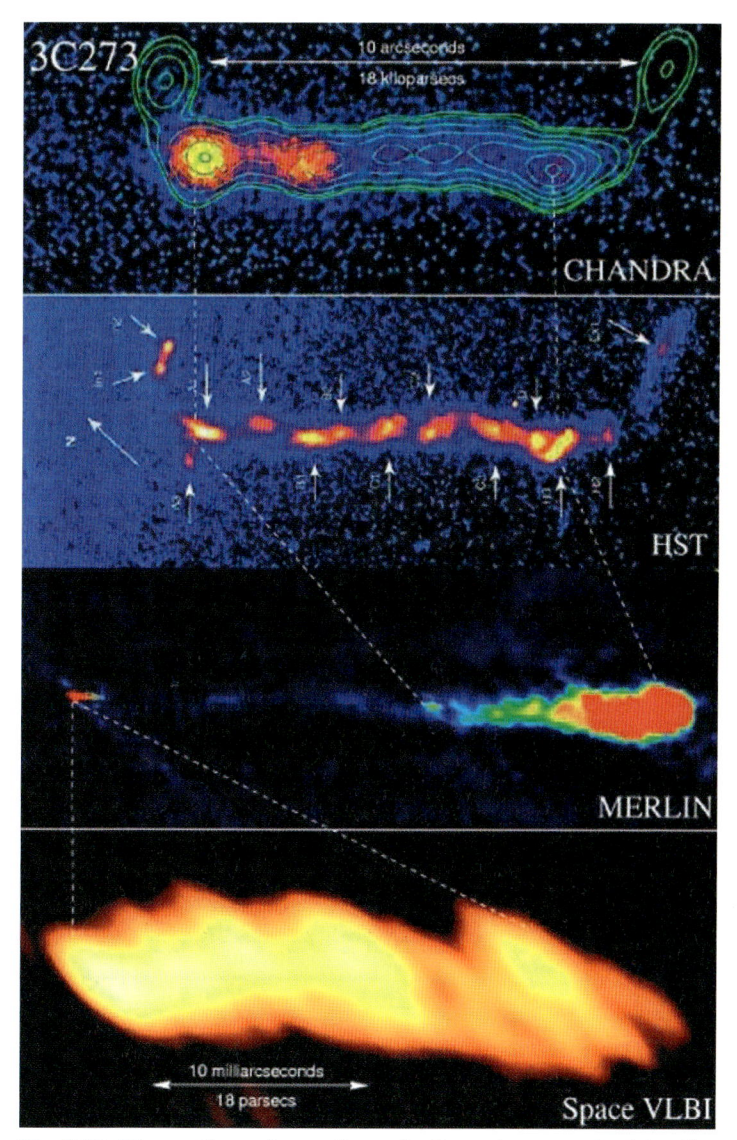

Fig. 2.33: Observations of an active galactic nucleus, emphasising the necessity for multispectral measurements. The spatial distribution of a jet from the region of quasar 3C273 is shown at different wavelengths (X-ray, optical, radio) and various resolutions (Chandra/HST/MERLIN/Space VLBI). Similar structures can be found at all wavelengths, including at the smallest scales, which can only be resolved using radio interferometry (VLBI). VLBI already achieves resolutions of 50 microarc seconds today; that is a world record for the highest image resolution attained in astronomy. (MPIfR)

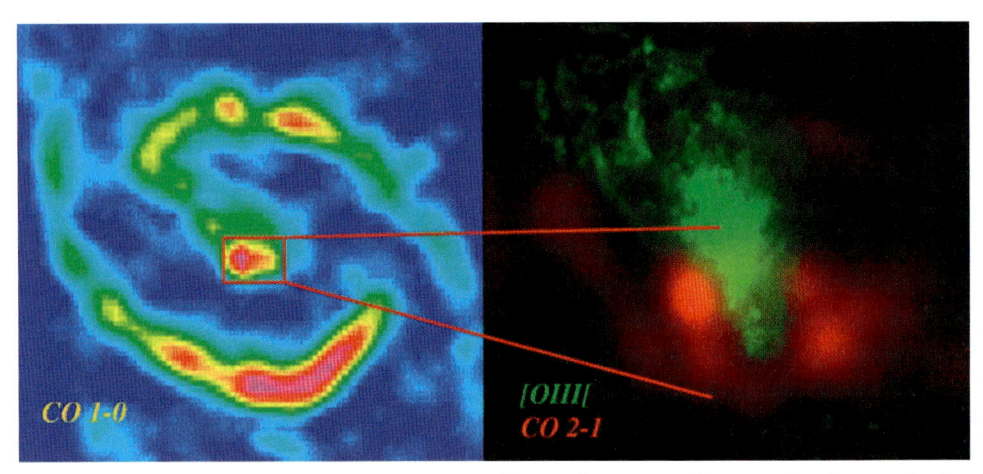

Fig. 2.34: Left: This image, taken with the IRAM interferometer, shows molecular gas in the active galaxy NGC 1068. It flows from two spiral arms along a "bar" and into the nucleus. Right blow-up: Molecular gas in the nucleus (red) envelops the conical region with ionised gas excited by the active nucleus (green). (MPE/IRAM)

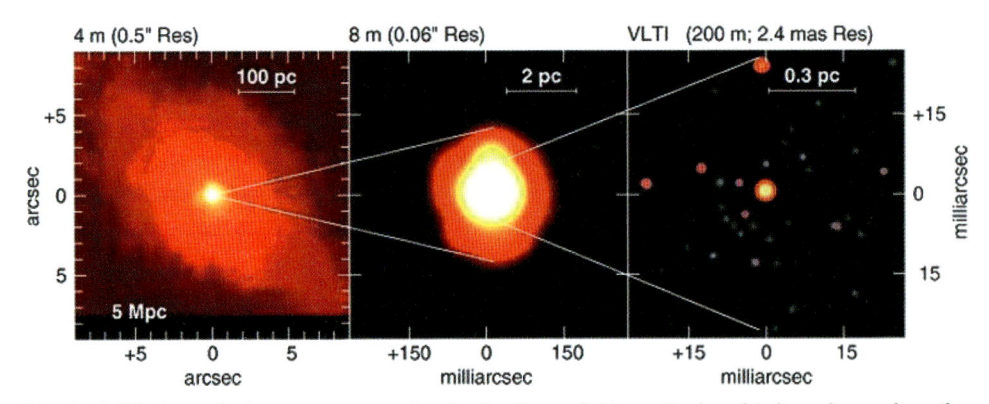

Fig. 2.35: High-resolution measurements of galactic nuclei in optical and infrared wavelengths. At the left is an image of a nearby (15 million light years) nucleus, taken at good "visibility". An unresolved, bright, active nucleus sits in the middle of a bar-shaped stellar distribution. Using adaptive optics on the VLT and LBT (centre), this nucleus can already be resolved in a compact central component of hot dust around the massive black hole, with a surrounding, more extensive cluster. Using interferometry methods individual bright stars can be recognised and investigated within this cluster (right). (MPE)

galactic nuclei. Obviously, stellar genesis can be promoted by the gas flux into the region around the nucleus.

Greatly enhanced stellar genesis and active nuclei also coexist in many of the very luminous galaxies. In recent years, German working groups have utilised infrared spectroscopy on the European space telescope ISO in order to penetrate the bothersome dust layers, and to demonstrate that the energy emitted by most of these objects is dominated by the newly created stars. The same question, which is also closely linked to the explanation of cosmic X-ray and infrared background radiation, will pose itself again in the next decade with regard to luminous galaxies with high redshifts, making observations with XMM-Newton, Chandra and Herschel, and later with XEUS, necessary.

Gamma observations

Active galactic nuclei are among the locations with the highest energy flux in the universe. Since numerous active galaxies have been observed in the gamma range with energies higher than 100 MeV, it is clear that a significant portion of the energy released when matter falls into a central black hole is emitted at the highest energies. Several active nuclei have been identified as sources of gamma radiation with energies in the TeV range. German working groups assume a leading role here with the HEGRA system.

In these cases, the radiation probably originates from tightly bundled particle beams, shooting into space in opposite directions from a central region. The particles (electrons and possibly their antiparticles, positrons) move at almost light speed in these jets. Observations of the Mrk 501 galaxy indicate that the high-energy radiation originates from electrons accelerated in a magnetic field. The H.E.S.S. and MAGIC instruments, currently being built, should enable data to be collected from more active galaxies in the future. They should also make it possible to discover more about particle acceleration processes, in particular in the immediate vicinity of a black hole.

Active galaxies and neutrinos

During the next decade, the new field of neutrino astronomy will not only generally open a brand new observation window, but also promises very tangible data to complement gamma-ray astronomy. High-energy neutrino detectors must be built extremely large in order to be sensitive enough. Although detection of individual sources will be difficult even with the planned ICECUBE neutrino detector, active nuclei should be identifiable as neutrino sources – and thus as heavy particle (hadron) accelerators – by superimposing the signals from several sources. An additional strong signal should also be the integral neutrino flux, generated by all active galaxies. However, an exact understanding of the background is necessary for this, because the identification of individual sources will not be possible due to the lack of directional information.

2.3 The matter cycle and stellar evolution

Scientific advances

- Insight into the distribution, physics and chemistry of the interstellar medium.
- Discovery of brown dwarfs.
- Discovery of microquasars.
- First ever observations of the sources of cosmic rays in the high-energy gamma range.
- Identification of gamma-ray bursts inside galaxies.
- Better understanding of the magnetic activity of the sun, the corona and the solar wind.
- Clarification of the physics of stellar winds.
- Probing the interior of the sun and other stars by helio- and asteroseismology.
- Identification of element abundances in supernova remnants.

Purpose and aims

- Structuring of the multiphase interstellar medium model and the interactions between stars and the interstellar medium.
- Dynamical models of the evolution of single and binary stars.
- Clarification of the nature of the precursor stars of Type Ia supernovae.
- Clarification of the accretion processes in neutron stars and black holes, as well as the origins of plasma jets.
- Numerical simulations of stellar explosions and merging compact objects.
- Detailed studies of gamma ray bursts at all wavelengths.
- Clarification of the magnetic activity of the sun and the stars.

The greatest proportion of the observable matter in galaxies, such as our Milky Way for example, exists in two very different states: on the one hand as a hot, dense plasma, for example in the interior of stars, and on the other as a cold, extremely attenuated gas interspersed with dust in the interstellar medium. There is a constant cycle between these two states. Because of its constant enrichment with heavy elements, the interstellar medium forms the foundation for the cosmic evolution of all matter, up to the complexity we see today (Figure 2.36). The

2.3 The matter cycle and stellar evolution

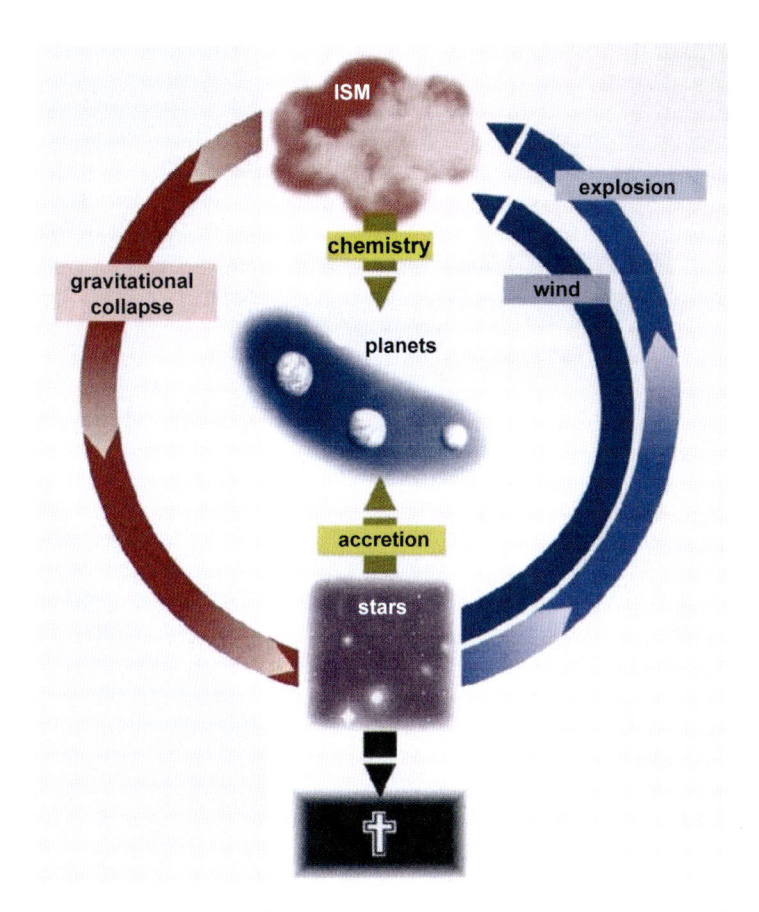

Fig. 2.36: The interstellar medium cycle: principal protagonists and important processes (University of Bonn).

medium that exists between the stars may only contribute a few percent to the total mass of our Milky Way, but it is the material from which the stars are made and which determines the evolution of the galaxy. However, the total mass of all galaxies is dominated by a halo of dark matter, whose gravitational potential determines the large-scale dynamics of the stars and gas of the universe.

Stars occur in groups of dozens to several hundred thousands of members in dense, cool clouds of interstellar matter. The young stars impact in different ways on the surrounding interstellar medium: by emitting radiation and by particle or stellar winds, which stream into space from their surfaces.

Only several million years after the formation of a stellar cluster, the more massive members already begin exploding

The cosmic cycle

as supernovae. As this happens they produce shock fronts that race through the interstellar medium. They partially destroy the dust particles, heat up the gas and ionize it. That is, the atoms are stripped of their electrons, creating a plasma of electrically charged particles. At the same time, large quantities of matter enriched with heavy elements, created in precursor stars by nucleosynthesis processes, are emitted into the interstellar medium. These later form clouds of molecules, and the matter cycle is closed.

The end products of stellar evolution – white dwarfs, neutron stars and black holes – remove more and more matter from the cycle. The gas reserves will therefore be exhausted at some time in the distant future and star formation will cease.

The stars play a central role in this circular process; that of motors. They are responsible for generating and distributing heavy elements. Their existence is a necessary requirement for the formation of solid bodies such as planets, and thus for biological evolution.

Formation of complex molecules

Without an external energy source, the interstellar gas would cool to temperatures around 10 K. In the interior of cool, dense clouds, where even the energetic radiation of hot stars cannot penetrate, conditions suitable for the formation of molecules prevail. By far the most common molecule in the universe is molecular hydrogen, H_2, which can only be directly detected in the infrared and far ultraviolet parts of the spectrum. It played the decisive role in the formation of the universe's first generation of stars. The formation of H_2 on the surface of dust particles is a necessary precursor to the formation of complex molecules. An increasing number of methods, some indirect, are continuously being developed, both to detect these molecules and to help understand the processes involved in stellar genesis, especially in cases of low element abundances, together with the complex structure and dynamics of molecular clouds (Figures 2.37 and 2.38). These processes are discussed in more detail in the following section.

2.3.1 The interstellar medium

An understanding of the matter cycle processes taking place in the interstellar medium forms the foundation for understanding many other fields of astrophysics: stellar and planetary genesis, the evolution of galaxies and the origins of the various types (see Section 2.2), and finally the chemical evolution of the universe. Due to the unusual conditions prevalent there,

2.3 The matter cycle and stellar evolution

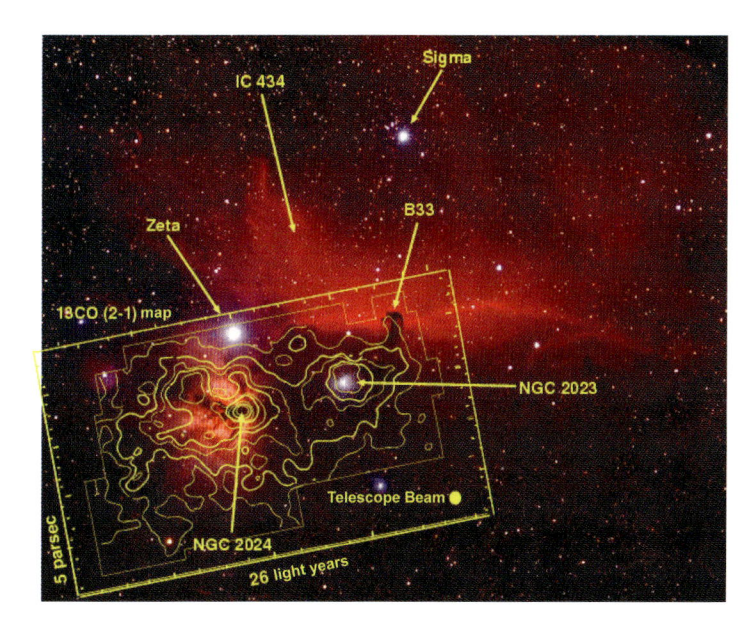

Fig. 2.37: The Orion B star formation region. The optical emission of some stars can be recognised. The contour lines show concentrations of carbon monoxide molecules (measured in the millimetre range), the locations of current star formation. (University of Cologne)

from a terrestrial point of view, the interstellar medium represents an ideal laboratory for studying the physics of highly attenuated plasmas, chemical processes under extreme conditions, atomic and molecular physics, but also solid-state physics (dust), as well as many other fields of the natural sciences.

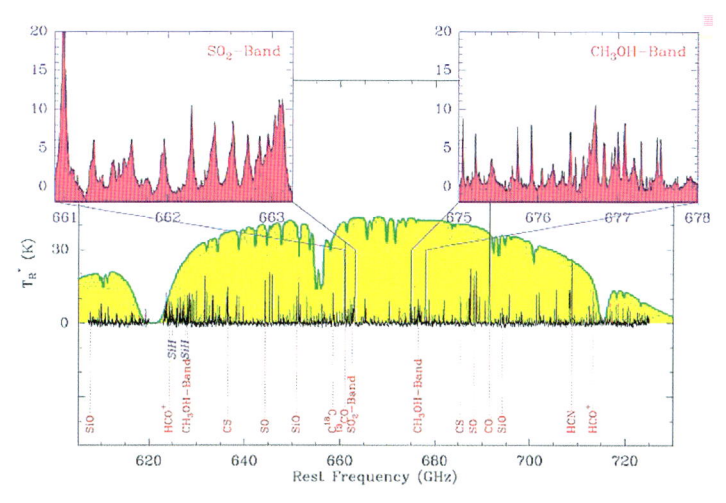

Fig. 2.38: Astrochemistry and astrobiology. The figure shows a spectral survey of the molecular emission lines in the atmospheric window at ~ 450 mm taken using high-resolution heterodyne spectroscopy on the CSO telescope on Hawaii (the yellow-green graph shows the atmospheric transmission on the 4,300 m high Mauna Kea). Using SOFIA and Herschel, similar surveys are possible for the entire infrared and submillimetre ranges. They provide a quantitative "chemical fingerprint", allowing indisputable conclusions to be drawn about the chemical evolution of interstellar or circumstellar clouds. (MPIfR and University of Cologne)

2 The scientific issues

Stars influence the interstellar medium

The stars exert an influence on the surrounding interstellar material. In their early phase, the numerous low-mass stars develop violent particle winds and bipolar streams (jets). In the late phase of their evolution, they cool enough that small solid-state particles form from the elements formed within them. Strong stellar winds drive these particles into the interstellar medium, leading to its enrichment with heavy elements and dust particles. These cool mass fluxes contribute extensively to the turbulent inner dynamics of molecular clouds. In contrast, massive stars are very hot and can therefore effectively heat up the surrounding gas, destroy molecules (dissociation) and ionize atoms. They also develop energetic stellar winds with expansion velocities of several 1000 km/s, which sweep the neighbourhood of the massive star clean and thus exercise a dynamic influence on the interstellar medium. The zone of ionized gas, with temperatures of around 10^4 K, that forms around hot stars can be observed and analysed thanks to its numerous emission lines in the optical, infrared and radio ranges (Figure 2.39).

Supernovae and the interstellar medium

Only several million years after the formation of a globular cluster, the more massive stars are already exploding. These supernovae are enormously important to the interstellar medium in a number of ways. They heat up the interstellar medium and are sources of cosmic radiation. However, they play their most important role in the matter cycle, as mentioned above, as sources of heavy elements. In the course of galactic evolution supernovae and stellar winds enrich the interstellar medium with heavy elements; these, in turn, determine how fast hot gas can cool again and be available once more for stellar genesis. The ROSAT all-sky survey provides a complete overview of the hot phase of the interstellar medium in the Milky Way (Figure 2.40).

X-ray spectroscope observations using XMM-Newton and Chandra have made it possible to directly observe the ejection of heavy elements in recent supernovae remnants. The densities of the most common elements, from oxygen to nickel, can now be directly determined. The supernovae remnants revealed in this way display heterogeneous compositions and distributions (Figure 2.41). The element abundances in the hot, interstellar medium of other galaxies can now also be determined. Similar observations of galaxies in the early universe, when the first heavy elements were formed, will be possible with XEUS.

2.3 The matter cycle and stellar evolution

Fig. 2.39: The Tarantula Nebula in the Large Magellanic Cloud, observed by the VLT. Young, hot stars have largely disintegrated the cloud in which they originated and illuminate the gas remnants in their neighbourhood. (ESO)

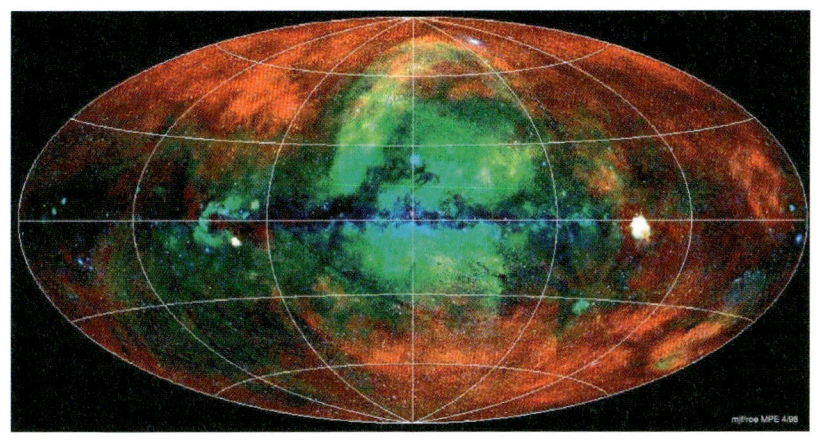

Fig. 2.40: The diffuse X-ray radiation in the 0.1–2.4 keV energy range, measured during the ROSAT all sky survey. The radiation is colour-coded according to energy level (red: low energy, blue: high energy). (MPE)

2 The scientific issues

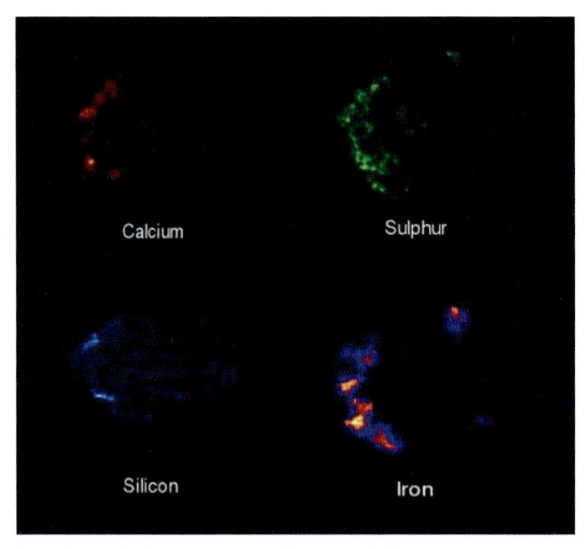

Fig. 2.41: XMM-Newton image of the supernova remnant Tycho, showing the distribution of the different elements. (MPE)

Turbulence and stellar genesis

The largest gas and dust clouds, known as giant molecular clouds, can contain as many as ten million solar masses. This makes them – with the exception of the black hole in the galactic core – the most massive objects in the Milky Way. Galactic rotation and the addition of kinetic energy by stellar winds and supernova explosions leads to highly turbulent motion in the interstellar medium. Low-mass clouds can thus grow to form massive molecular cloud complexes. Simultaneously, the inner turbulence leads to the formation of structures and substructures on all size scales. As described in Section 3.4, these processes are crucial to understanding stellar genesis.

Cycle of gas through the Milky Way's halo

Supernovae are capable of propelling gas out of the plane of a disk galaxy. This hot matter can be detected in our Milky Way and other spiral galaxies in X-ray and far ultraviolet observations. It is located in the outer zones, the so-called halo. The question now is: what happens to this gas and in what ways does it influence the evolution of a galaxy?

If the galaxy can hold the gas in the halo with its gravity, it will unavoidably (for lack of energy supply) cool and fall back into the disk plane. In fact, large clouds of electrically neutral gas can be observed, in several spectral ranges, approaching the galactic disk at very high velocities. The interstellar gas that exists in the Milky Way's disk is mixed and chemically homogenised by the hot gas streaming out into the halo and subsequently falling back at different locations in the

2.3 The matter cycle and stellar evolution

shape of high-velocity clouds. This mixing also reveals itself indirectly as fluctuations in the amount of heavy elements in stars in the same age group. At the same time, a systematic change in the level of heavy elements can be recognised in stars from the inner to the outer zones of the Milky Way. This phenomenon, too, is a measure of the efficiency of mixing in the interstellar medium.

High-velocity clouds

This interpretation of the high-velocity clouds observed in the 21-cm line of neutral hydrogen is not undisputed. A small proportion of these clouds are probably caused by gas being pulled out of the Magellanic Clouds by the Milky Way's tidal field. This can be derived from their positions in the sky. If parts of these high-velocity clouds were in fact located at much greater distances, they would have to be interpreted as gaseous components of the low-mass galaxies predicted by the theory of cosmic structure formation, in which stellar formation was, however, subdued. These galaxies would then be members of the Local Group and would be dominated by dark matter, similar to massive galaxies. More sensitive radio observations in the future will tell us whether similar high-velocity clouds are located around other spiral galaxies, and will be capable of directly measuring their separations. If this explanation proves to be true for high-velocity clouds, yet another prediction of the cosmological model would be confirmed. This is why this question is being very closely followed.

Observations in the entire spectral range

The diversity of thermodynamic conditions and the very different processes acting in the interstellar medium produce a wide range of observable phenomena. Researching the interstellar medium, it can be seen, requires observing and measuring equipment that covers practically the entire electromagnetic spectrum: in the gamma-ray range, we can observe the interaction of cosmic rays with dense molecular clouds and, using a spectroscope, the distribution of heavy elements generated during supernova explosions; in the X-ray range, the radiation emanating from hot gases; in the far ultraviolet, the absorption lines of molecular hydrogen (H_2) and heavily ionized oxygen atoms (O^{5+}); in the ultraviolet, the absorption lines of many metals and of carbon monoxide (CO); in the optical range, the classical absorption and emission lines from regions containing ionized gas and diffuse absorption bands; in the near- and mid-infrared, the emission of H_2 and of ionized gas, as well as absorption structures in the interstellar dust; in the far infrared, the emission of dust and important emission lines of cool and warm gases; in the *submillimetre, millimetre* and *centimetre wavelengths*, the emission lines of numerous molecules and of atomic hydrogen; as well as the continuum emission of hot, ionized gas and synchrotron radiation in the radio range.

Observation opportunities, today and tomorrow

Very good observation opportunities exist for many of the wavelengths mentioned, for example with the Franco-German IRAM observatory in the millimetre and submillimetre ranges, with ESO's VLT in the visible and near infrared, or with the XMM-Newton (ESA) and Chandra (NASA) X-ray telescopes. In terms of the detailed investigation of local phenomena in the Milky Way, ALMA will provide new contributions in the millimetre and submillimetre ranges, thanks to its high angular resolution. A gap has opened up following the European space missions IRAS and ISO, which will first be bridged by the American SIRTF Mission and then closed by Herschel and SOFIA. In the ultraviolet, the HST provides a flexible, and therefore heavily overbooked, telescope. Its tasks will be taken over by the World Space Observatory/UV after it is switched off in 2010. The Franco-American FUSE satellite operates in the far ultraviolet. In the gamma-ray range, INTEGRAL will provide precise heavy element distribution maps and prove the existence of more nuclides. Similar observations of galaxies in the early universe will be possible with XEUS. Finally, the structure of the interstellar medium can be surveyed by utilising the long wave radiation from pulsars, propagation processes such as interstellar dispersion, Faraday rotation, dispersion and scintillation. The Effelsberg today, and later the SKA, will be used to achieve this.

2.3.2 Cosmic rays

Large energy range

An unusual component of the interstellar medium is represented by cosmic rays. They consist of particles whose energies cover many orders of magnitude from several million electron volts (eV) up to more than 10^{20} eV. In extreme cases, a single atomic nucleus possesses kinetic energy comparable to that of a tennis ball at service! The energy distribution is not thermal and contains no indication of a characteristic energy or temperature scale, such as that found in thermal radiation. In addition, the energy density of cosmic rays is comparable to that of the turbulence in the interstellar gas or the energy density of starlight. This indicates that cosmic rays are not an isolated phenomena, but instead are in a kind of equilibrium with other forms of energy.

2.3　The matter cycle and stellar evolution

Possible accelerator sources

The search for sources and acceleration mechanisms for high-energy cosmic rays is one of the fundamental topics of high-energy astrophysics. It is currently assumed that particles are accelerated up to 10^{15} eV in shock fronts emanating from supernova explosions. However, there is no proof of this. The even more energetic particles may originate in active galactic nuclei.

An indirect key to understanding the generating mechanisms and propagation of cosmic rays is their elemental composition. In the lower energy ranges, the mean residence time in the Milky Way can be derived from isotopic ratios. The composition of cosmic rays in the 10^{15} eV range in particular, has been and remains the object of intense investigations, for example using the KASKADE and HEGRA air shower arrays. The energy distribution displays a bend in this region (called a knee); we can only speculate on its origin.

First sky map

Because the charged particles of the cosmic rays are deflected by the galactic magnetic field, their routes cannot be traced back to their origins. However, the particles do interact with matter and radiation fields in the universe. High-energy gamma quanta are produced, allowing the sources and propagation of cosmic rays to be identified. Mapping the diffuse gamma radiation in the energy range covering up to several GeV was first achieved in the mid-1990s with the EGRET instrument on board the CGRO gamma ray observatory. German researchers were involved in building the observatory. The map shows strong emissions from the plane of the Milky Way. They originate in part from individual sources. A diffuse radiation component also exists and is generally distributed homogeneously throughout the Milky Way. It is probably the result of the interaction of cosmic rays with interstellar matter (Figure 2.32, left).

Future gamma astronomy instruments in the several GeV to TeV range, such as GLAST, H.E.S.S. and MAGIC, will allow detailed modelling of the distribution of cosmic rays in the Milky Way. The sources of cosmic rays, for example the suspected supernova remnants, will be made visible by the TeV gamma quanta generated within them.

2.3.3 The nearest star: the sun

There are more than one hundred million stars in the Milky Way, yet only one of them is close enough to allow direct observations of physical processes on a number of scales: our sun. (Magneto-) hydrodynamic processes occur in and on the sun at very different spatial scales, ranging from a few kilometres up to dimensions as large as the sun's diameter. The scales of temporal variations range from seconds and minutes for a flare eruption, via the years-long scale of the solar activity cycle, up to magnetic activity, and thus luminosity, fluctuations extending over centuries to millennia.

Helioseismology

The sun is a giant ball of gas that vibrates at diverse frequencies. Today, analysis of these intrinsic vibrations is an effective tool for probing its inner structure. The procedure is reminiscent of seismology on Earth and is consequently called helioseismology. Precise measurement of the vibration frequencies over many years allows experimental determination of state variables as a function of the radius, and comparisons to solar models. This gives an exact picture of the depth- and width-related rotation in the sun's interior. For example, these studies have shown that the width-related (differential) rotation speed observed on the sun's surface is also prevalent in the interior (into the convection zone).

Moreover, helioseismology and solar neutrino research complement each other perfectly. The latter only allows statements on the physical state of the immediate centre of the sun, whilst it is exactly this region that is difficult to observe using helioseismological methods.

Interactions between convective flow and magnetic fields

Up to now, dynamic interactions between convective flow and the magnetic field in a plasma can only be studied on the sun at its characteristic length and time scales. In the photosphere and chromosphere the sun's magnetism manifests itself as sunspots, pores and magnetic nodes, right down to small-scale flow elements. The spatial dimensions of magnetic phenomena extend across several orders of magnitude, from more than 100,000 km to far less than 100 km. The smallest scales are far below the limit of resolution dictated by diffraction of even the largest of today's solar telescopes.

Variations and their influence on Earth's climate

Recently it has become obvious that the variation in solar magnetism correlates to a fluctuation in energy output (the so-called solar constant). This effect is small in the visible and infrared ranges: the long-term fluctuation is around 0,1%. In the extreme UV and X-ray ranges, in contrast, the radiation varies during each sunspot cycle by around an order of magnitude. The exact causes of these fluctuations and the processes involved are currently the subjects of intense research, as are their influence on the region in the solar system dominated by the solar wind (heliosphere), and thus ultimately on Earth's climate.

Origin of the magnetic field

All solar activity phenomena vary on an 11 or 22 year cycle. The regular reversals in the magnetic field at each cycle change are interpreted as evidence for a magnetic dynamo. Despite enormous theoretical efforts, the modus operandi of the solar dynamo is still unclear, in particular in terms of the significance of magnetic flux tubes. Moreover, theory only allows very few quantitative predictions for the sun, and almost none for other stars.

The solar-stellar relationship

One way of finding out more about the sun's inner magnetic field is to observe sunlike stars. If objects are selected that differ from the sun in either their size, mass, chemical composition or age, the influence of these variables on magnetic activity can be identified (solar-stellar relationship). This makes it possible to draw conclusions on the sun's activity in its youth, for example, and on how often and under what conditions activity cycles occur in stars. Previous advances made in unravelling the solar-stellar relationship were the result of improved observation opportunities. During the next decade we will also have the opportunity to acquire high-resolution optical spectra from stars using the large telescopes, or with observations in the UV and X-ray range. High-precision photometry will in future allow the study of oscillations in nearby, sun-like stars. Tighter interlinking to theoretical models will then be a matter of urgency. Observations over long time scales such as the 11 year sunspot cycle require the use of specialised telescopes with mirrors up to three metres in diameter. Completely automatic operation will be an important element of their success.

Corona and solar wind

Studying the sun has revealed to us the complexity of the properties of outer stellar atmospheres and their coronas, which we can only see with the naked eye as a bright, encircling diadem during a total eclipse. The corona represents the transition zone to interplanetary space. The sun's magnetic field structures its atmosphere and links the individual layers. The atmosphere can therefore only be understood as a system coupling radiation, plasma and the magnetic field.

Fig. 2.42: A coronal mass ejection from the solar corona with a twisted magnetic field, taken by the LASCO instrument on the SOHO space observatory. If these particle streams impact Earth's magnetic field at speeds of more than 1,000 km/s, they initiate magnetic storms and induce the polar lights. (ESA/NASA/MPAe)

The solar wind – and the heliosphere generated by it – is the only stellar wind in which particles and fields can be measured in-situ and where the corresponding plasma processes can be investigated by experiments in space (Figure 2.42). The knowledge gained from satellites and space probes is of exemplary character for understanding other, very much more remote plasmas in the cosmos, and the origin of the winds of other stars. Here, too, the question of the influence the solar wind exerts on the Earth and its magnetic field is at the centre of our interest.

2.3.4 The stars

Stars are the most investigated and theoretically best understood heavenly bodies. The physics of stellar structures and the evolution of stellar atmospheres has been a focal point of astronomical research for decades, especially in Germany. Enormous progress has been made here. Large areas of stellar physics are now understood, which is why stars are exceptionally well suited to studying physical processes. More recent

2.3 The matter cycle and stellar evolution

observations have raised new questions, the answers to which promise important insights into physics.

Stars as plasma physics laboratories

An understanding of the physical processes taking place in the outer envelopes of stars is critically important to the whole of astrophysics. Plasma processes, such as coronal heating to more than one million degrees, short-circuits in magnetic fields (reconnection), phase transitions, stellar wind acceleration and acceleration of particles to relativistic energies, are very probably prototypes for processes that are widespread in the cosmos. They can be found in similar forms in the diffuse interstellar medium, in the disks around neutron stars and black holes, in galaxy clusters and in other objects.

Brown dwarfs

One of the outstanding discoveries of the last decade was the long-predicted existence of brown dwarfs (Figure 2.43). These are cool, very faint, low-mass objects that do not generate appreciable amounts of thermonuclear energy. In terms of mass, they are situated between stars and planets. Previous observations have confirmed predicted properties very well. However, at the same time, the new data emphasise the necessity for theoretical models of both the atmospheres and the inner structure and evolution of brown dwarfs. These models currently fall behind observations.

"Weather" on brown dwarfs

At the relatively low temperatures prevalent in the dense atmospheres of brown dwarfs, chemical and physical non-equilibrium processes play a central role. These processes cannot be described by classical atmospheric models. In particular, observations have shown that dust particles form in these atmospheres. There are even indications of cloud and wind formation, so that we can actually speak of changing "weather conditions" on these heavenly bodies. In the future, these phenomena will require considerably more detailed descriptions of their atmospheric physics, which will also place greater demands on numerical models.

Stellar winds

In the field of hot stars, models allow reliable identification of both the physical parameters and the chemical composition of static and expanding stellar atmospheres. Refined standard atmosphere models exist for analysing stars with static, not excessively hot, atmospheres. However, in certain phases of their evolution stars also lose a large percentage of their mass, sometimes even the majority, due to continuous stellar winds or episodic gas ejections. Some of the physical mechanisms involved have recently been understood. The winds from hot

2 The scientific issues

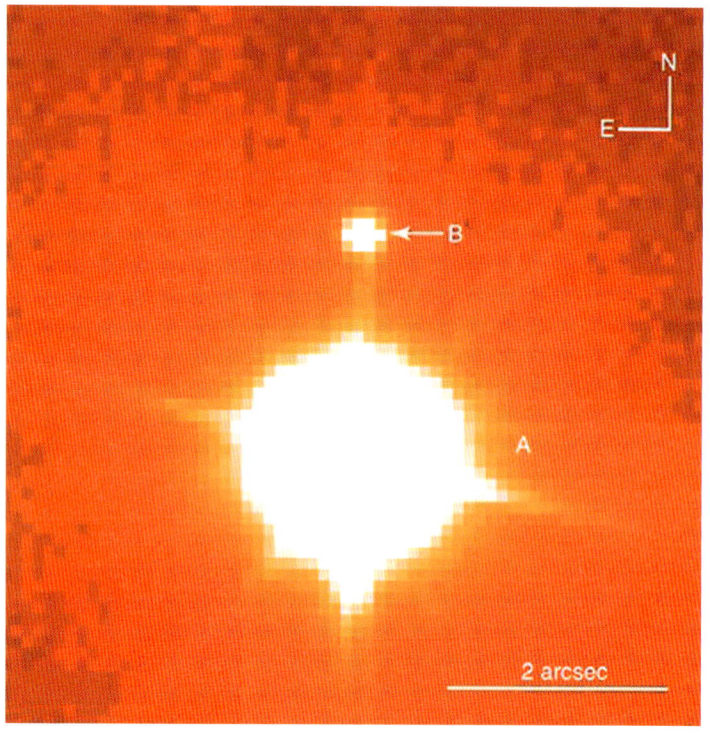

Fig. 2.43: A brown dwarf companion of the star TWA 5. (ESO)

stars can be quantitatively explained by the effects of radiation pressure on atoms (Figure 2.44). Self-consistent dynamic models are now only lacking for very dense winds (from Wolf-Rayet stars). Here, a heterogeneous stellar wind structure was discovered, especially in the X-ray range, for which no explanation is yet forthcoming.

The mechanism driving the winds in cool giant stars is not very well understood. In the extreme final phases of giant stars the wind may be driven by the radiation pressure on the dust formed inside the cool envelopes. Pulsations of the whole star may also play a role here. Various observation variables such as light curves, intensity and line profiles, and the radial concentrations of various molecules can already be quantitatively described based on consistent models.

Close binary star systems

Nearby binary stars also make very interesting objects of study. They orbit one another at such close distances that matter is drawn from one to the other. During this process the gas often initially collects in a disk. Here, it loses energy before falling into the star. The second star accumulates matter,

2.3 The matter cycle and stellar evolution

> **Box 2.3: Stellar physics: a traditional domain of German astronomy**
>
> The structure and evolution of stars is one of the best understood fields of astrophysics. German astronomers have taken up leading positions worldwide in this field since the beginnings of modern astrophysics. Stars are excellent laboratories for studying (astro)physical processes, for example radiation processes and the interaction of matter and magnetic fields. Both hot and late, cold stars, for example, are essential motors for the cosmic matter cycle. These stars have enormous luminosities, millions of times greater than the sun. The radiated light alone exercises such pressure that huge quantities of matter are accelerated to high speeds and propelled away from the star – the stars partially disintegrate. Both the radiation-driven winds of hot stars and the pulsation- and dust-driven winds of colder stars have been extensively researched by German astrophysical institutes. Observational data was provided by the ESO and Calar Alto telescopes; the space-based UV spectrographs came from the IUE and HST, and from ROSAT in the X-ray range. Physically and numerically complex though they are, the spectra required for interpretation of the stellar atmosphere models were successfully developed. They now allow the required parameters to be reliably determined. The evolutionary paths of both the hot, massive, and the cool, less massive stars, have been at least partially clarified.

which is why we speak of accretion disks. Instabilities in the gas flow often occur in such systems and flares are related to this. Studying these cataclysmic variables allows the structure of accretion disks, which are too small to be directly observed, to be determined. Knowledge gained both in theory and from observations can be transferred to other objects such as X-ray binaries or the nuclei of active galaxies. The investigation of hot accretion plasmas using the new X-ray satellites Chandra and XMM-Newton presents an enormous challenge. XEUS may provide considerably more data in the future.

The importance of magnetic properties has been impressively confirmed for a number of stars thanks to instrumentation advances. For example, during the last decade it has been possible to detect cyclic variations in the strength of calcium emissions. As we know from the sun, this emission line is a

Magnetically active stars

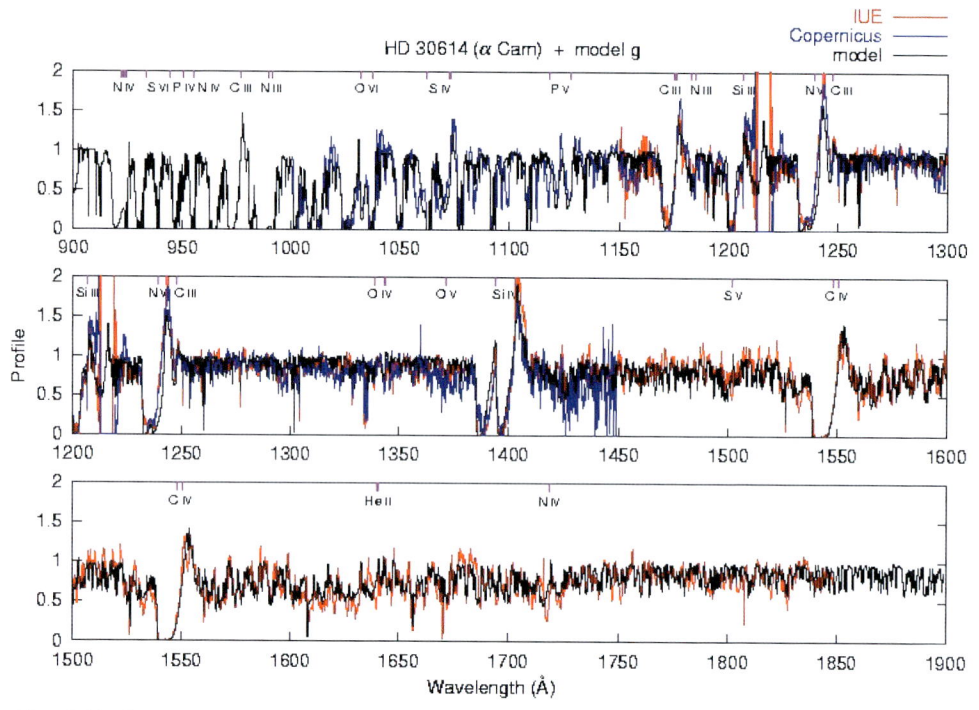

Fig. 2.44: Quantitative stellar spectroscopy. Detailed comparison of the high-resolution UV spectrum of an O-type star (α Cam, red and blue) and a non-equilibrium model of the stellar atmosphere. (University of Munich)

measure of the magnetic activity of a star. Whilst observing the variations in the calcium lines of other stars, variations similar to the 11-year solar cycle were found. Using an ingenious investigation method (Doppler tomography), it was also possible to map magnetic field structures on stellar surfaces, even though they could not be resolved (Figure 2.45). X-ray observations have shown that coronas are universal among sun-like stars. Magnetically driven particle winds and the occurrence of jets in both young stars and active galaxies alike are regarded as being closely linked to the magnetic activity of stars. However, further observations are necessary in order to completely understand these phenomena.

2.3 The matter cycle and stellar evolution

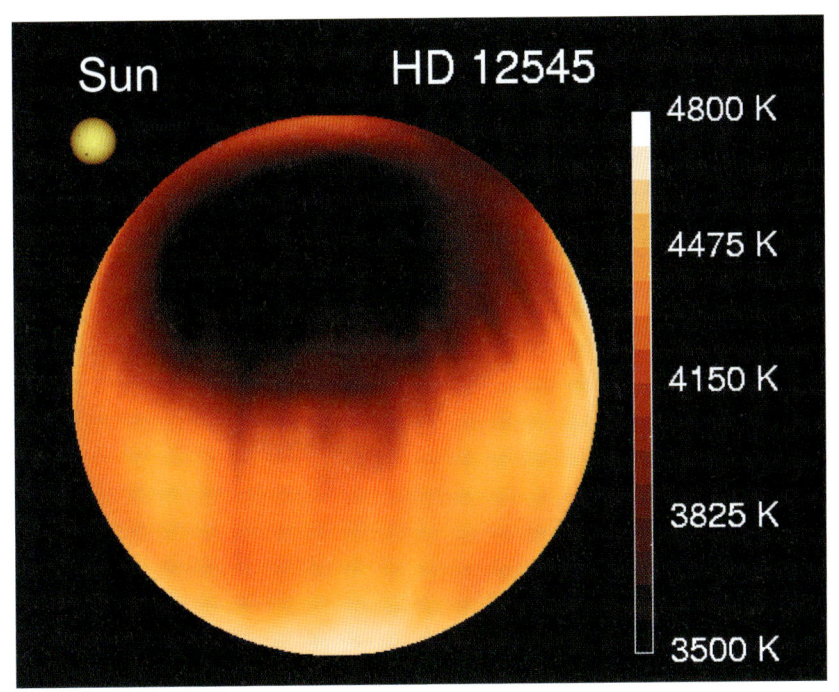

Fig. 2.45: Surface mapping of the star HD 12545 with the aid of Doppler tomography. (AIP)

2.3.5 Stars as chemical factories and the engines of the matter cycle

Stellar evolution

The theory of stellar evolution occupies a key position if we want to understand the enrichment of the interstellar medium with heavy elements. This includes, on the one hand, physically describing element enrichment in stellar atmospheres and stellar particle winds, and describing mass losses as a function of the stellar parameters. On the other hand, knowledge of all possible nuclear fusion processes and reaction rates is important, both for the conditions prevalent in normal stars, and for exceptional conditions, for example as they exist when gas streams impact compact stars.

Particle winds enrich the interstellar medium with elements

The size, temperature and luminosity of low-mass stars like the sun remain practically constant for several billion years after their relatively brief birth phase, which lasts only a few million years. When the hydrogen reserves in their interiors are finally consumed, the pressure and temperature increases. Helium is then converted to carbon and oxygen. As a consequence of this altered inner structure, the star expands to form a giant. The second giant stage is particularly important for the cosmic matter cycle, because the star now loses a great deal of matter. By this process, these stars influence the chemical evolution of the Milky Way.

Great progress has been made in the consistent inclusion of matter losses in the analyses of stellar evolution. Today, models can describe the "superwind" characteristics of planetary nebula (Figure 1.6) as easily as they can the shell-like envelopes generated by carbon-rich stars (Figure 2.46). The latter are formed by very violent, but brief surges in mass loss. Envelope physics, that is the interaction between hydrodynamics, radiation transport and chemistry, therefore plays an important role in the later evolutionary phases of a star.

Dynamic processes in stellar interiors, such as convection or thermal pulses, play a decisive role in stellar evolution. The current standard stellar evolution theory has not yet been able to reproduce the results of spectral analyses. It is therefore absolutely necessary in future to consider the dynamic phenomena involved in stellar evolution.

In close binary systems the evolutionary paths of both members are interlinked. If one component develops into a giant, matter will flow to the companion. It is possible for a common envelope to form around both stars. This phase has a

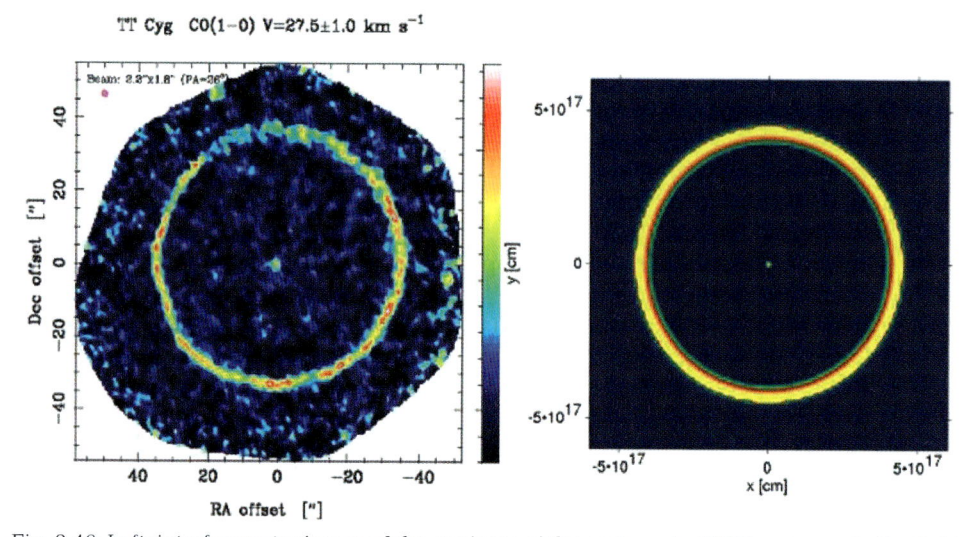

Fig. 2.46: Left: interferometer image of the environs of the carbon star TT Cygni seen in the light of the CO molecule. Right: hydrodynamic model of the particle wind. If a critical mass loss rate is reached, the wind speed leaps abruptly to a higher value. Matter is pushed into a shell and compressed in the resulting interaction zone. (AIP/TUB)

2.3 The matter cycle and stellar evolution

critical influence on the future evolution of the binary, but is so far only poorly understood.

Dust formation in circumstellar envelopes

Gas and dust envelopes, especially around low-mass stars, are ideal objects for studying the prevalent chemical processes. The chemical structure can be studied on the basis of infrared and radio observations of molecules. It has been known for some time that dust also forms in the envelopes. During this process, which occurs in stages via chemical reactions in the gas phase, matter goes through a series of steps of increasing complexity. Fundamentally different chemical relationships occur depending on the element composition: in carbon-rich element mixtures, organic chemistry prevails and carbon and hydrogen primarily contribute to dust formation. In contrast, inorganic compounds form in oxygen-rich environments, dominated by various silicates.

ALMA's excellent angular resolution capacity and great sensitivity will make it possible to model the spatial distribution of diverse molecules even within the dust formation zones of giant stars. Direct measurements of the frequencies of many molecules as a function of the distance from the central giant star will provide crucial information on dust formation.

Galactic evolution

One of the central aims of astrophysics now consists of quantifying as precisely as possible the contributions of the various stellar types to the enrichment of matter within the Milky Way. The data used is based on both observations and models. This helps to determine the chemical evolution of the galaxy. The galactic evolution models can utilise the properties of stellar components derived directly from detailed stellar wind and supernova models, instead of relying on approximate equations or estimates. This makes it possible to bridge the gap between stellar and galactic astrophysics.

In the future it will also be possible to precisely determine the physical properties of massive stars in the relatively nearby galaxies of the Local Group, including their winds and exact element abundances. Hot, low-mass stellar populations in our neighbouring galaxies, which dominate the UV light from some of them, can be studied for the first time. It will thus be possible to investigate stellar evolution as a function of the occurrence of heavy elements, for example. This has considerable consequences for our understanding of the spectra of remote galaxies and therefore for galactic evolution.

2.3.6 The final stages of stellar evolution

In a manner of speaking, matter is "buried" for all eternity in the final evolutionary stages of some stars. These include white dwarfs, neutron stars and black holes. They represent sinks in the matter cycle. In addition, matter in the interior of these bodies finds itself in extreme physical states that cannot be reproduced in laboratories on Earth. The spectacle provided to us by stars just before their final stage of existence is particularly impressive.

White dwarfs

Stars with an initial mass of less than around eight solar masses – including our sun – discard their envelopes to form a planetary nebula and end up as white dwarfs. At first they are very hot, but then slowly cool because nuclear fusion is no longer taking place in their cores. The theory of the genesis and cooling of white dwarfs is very well developed. A classical mass-radius relationship, discovered by S. Chandrasekhar, has nevertheless proven to be difficult to confirm empirically. This will change with the future launch of the astrometry satellite GAIA, as considerably more precise distance data will be available for numerous white dwarfs.

In recent years it has also been possible to apply helioseismological methods (see above) to pulsating stars. Astero-seismology has proved itself spectacularly useful in the case of white dwarfs, where stellar mass, chemical stratification and rotational velocity have been derived with great precision. Current cooling theories incorporate such exotic processes as cooling by neutrinos or by the crystallisation of matter. They will also be tested in the future on pulsating white dwarfs. Model atmospheres for hot, white dwarfs, which consistently take non-equilibrium thermodynamics, diffusion and the influence of heavy elements into consideration, for the first time quantitatively described certain observational characteristics (for example, very strong attenuation of the flux in the X-ray and extreme UV ranges). This is due to the absorption behaviour of the heavy elements. This helps us understand the relatively small number of white dwarfs discovered in the ROSAT all sky survey. Cool, white dwarfs are bluer than previously assumed due to the influence of various molecules. This is an important finding in terms of the future search for these objects. Most recently, white dwarfs were discovered in a halo population and there has been some speculation as to whether these white dwarfs can be regarded as an explanation for the observed microgravitational lens effects in the direction of the Magellanic Clouds. The cooling theory and atmospheric

2.3 The matter cycle and stellar evolution

models also represent central pillars for the luminosity function of white dwarfs, which can be used to determine the age of the stellar population in the neighbourhood of the sun.

If the mass of a white dwarf is increased to a critical value (known as the Chandrasekhar limit) by the influx of matter or by merging with a close companion in a binary system, the star is expected to become unstable and explode. As we understand it today, this is the process that we observe as a Type Ia supernova. The first candidates for the precursor systems of these stellar explosions have recently been found with the discovery of "super soft X-ray sources" by ROSAT, and a sufficiently massive, close binary.

Type Ia supernovae

Despite these advances, our knowledge of precursor stars is relatively small. This is in contrast to the great importance of the Type Ia supernovae, both as the main producers of iron and as cosmological standard candles for distance determination. Clarification can come from additional observations of candidate systems or by identifying remnant components. The theory of the evolution of close binaries and models of explosion process must be developed further in order to lead to an understanding of the Type Ia supernovae phenomenon.

Collecting a sufficiently large number of, and therefore dependable, samples is important for understanding the evolution of binary systems. Large search programmes, in which more than 1000 white dwarfs are being investigated for radial velocity variations with the help of the VLT, have already begun and have even found a further interesting candidate. Follow-up observations by the Sloan Digital Sky Survey may also provide relevant results.

Stars with initial masses of around eight solar masses are massive enough to burn even heavy elements in their cores. Towards the end of its life the star develops an onion-like structure, with heavier elements towards the centre. If the nuclear fusion chain advances as far as iron, no further energy can be generated; the intrinsic gravitation can no longer be withstood by the internal combustion pressure and the star collapses. The processes involved in this nuclear collapse are linked to extreme states of matter; an unstable interim condition develops, leading to a violent explosion that tears the star asunder. One result is a Type II supernova, which ejects the majority of its matter into the interstellar medium. The other result is a neutron star or, for the most massive stars, a black hole.

Type II supernovae

Investigations of this type of supernovae were intensified worldwide at the end of the 1980s after the discovery of Supernova 1987A in the Large Magellanic Cloud (Figure 2.47). It was close enough to be identified with the naked eye. Refinement of theoretical investigations and more precise determination of fast and slow nuclear reactions are also important for supernova explosion research. The amount of heavy elements ejected into the interstellar medium depends greatly on the details of the explosion processes. These processes need to be described more precisely. The heavy elements generated in supernovae contribute considerably to the cooling of the interstellar medium. Only under these conditions is particulate growth possible. The processes acting during a supernova explosion are thus extremely important in terms of the effects of the matter cycle on the interstellar medium.

Neutron stars

The matter in the interior of neutron stars is as densely packed as an atomic nucleus. One cubic centimetre of the matter contained in a neutron star would weigh a billion tonnes on Earth. It is also assumed that it is at least partially in a condition of suprafluidity. Matter displays no friction in this state. It does not

Fig. 2.47: Due to its relative proximity to Earth, Supernova 1987A in the Large Magellanic Cloud was a prize example for this field of research. (NASA/STScI/ESA)

exist naturally on Earth and can only be created in laboratory experiments. More information on this exotic state of matter, represented by the so-called equation of state, is therefore also expected from the study of neutron stars.

Neutron stars were discovered more than thirty years ago and were then known as radio pulsars. Their pulsating radiation, with periods as short as only a few milliseconds, identified them as highly compact objects, rapidly rotating and with magnetic fields up to a trillion times stronger than Earth's. Due to their extremely regular rotation periods, pulsars are exceedingly useful lighthouses in terms of studying the interstellar medium in the Milky Way. The radiation mechanism in the extreme magnetic fields of radio pulsars is still poorly understood even after more than thirty years of intense research. Significant advances are expected in the future thanks to the enormous sensitivity of SKA, and further observations aided by large individual instruments such as Effelsberg, in combination with detailed plasma physics investigations.

The direct thermal radiation of neutron stars was not detected until the arrival of ROSAT. The radiation originates in the star's atmosphere and allows the temperature to be determined. Neutron star cooling is primarily controlled by the generally unknown properties of the very dense matter in their interiors. It is hoped that temperature measurements will provide valuable indicators for the state of matter there.

According to current models, neutron stars that formed more than one million years ago must have already cooled to less than 100,000 K and would eventually become totally invisible, if it were not for mechanisms that slow cooling by generating further heat. One possibility is energy dissipation by the friction of a rapidly rotating suprafluid stellar interior against the outer, solid crust of the neutron star. Temperature determination in neutron stars would allow conclusions on core-crust interactions, as well as on their magnetic properties.

Models of neutron star atmospheres are therefore especially important. Innovative methods must be applied to very hot and dense plasmas in very strong magnetic fields. This presents numerous difficulties. On the other hand, these objects represent unique laboratories for studying plasmas with extreme magnetic fields (Figure 1.7). These models may be tested in the near future, based on high-resolution X-ray spectra.

Stellar black holes

If the collapsing star that remains after a supernova is larger than around three solar masses, it collapses further to form a (stellar) black hole. Because no electromagnetic radiation can escape from these objects, all information on them comes from their environment. The observed X-ray radiation, for example, originates in the accretion disk surrounding the black hole, from which it sucks up material. The physical processes that occur are very similar to those in active galaxies. Because of the general correlation between the time scales in question and the mass of a black hole, the black holes in our galaxy allow these processes to be investigated for a variety of characteristic time scales. This is very difficult when applied to massive black holes in galactic centres.

One elementary discovery at the beginning of the 1990s was that of the so-called microquasars. These are black holes that eject two tightly bundled gas jets. Using radio interferometry, it was shown that these jets move almost at light speed, similar to the much brighter quasars. It is therefore possible to study the formation of jets at the smallest scales on these astro-physical objects, as well as the interactions between the jets and the accretion disk.

More than 30 years after their discovery, many fundamental properties of galactic black holes remain a mystery, despite these advances. For example, there is no observational method available today that will allow us to decide, from the spectrum alone, whether a compact, matter-accumulating object is a black hole or a neutron star. The physics involved in the generation of radiation, and the processes operating to form the hot electron plasma and the jet, remain largely unclear. High-resolution measurements of X-ray spectra are vital in order to understand these physical processes. Precise radio observations will show the physical relationships between the accretion disk and the plasma jets.

Moreover, compact objects in different galaxies can also be observed and studied using these instruments. For the first time, this will allow these populations to be compared to that of the Milky Way.

Gamma ray bursts Among the most mysterious aspects of modern astrophysics are the so-called gamma ray bursts. They are short bursts that appear in the sky, on average once a day, completely without warning, and generally lasting only a few seconds. These "flashes" of intense gamma radiation were first noticed in the 1960s by military satellites designed to monitor atomic weapons tests. For a long time the nature of these sources was completely unknown; speculation ranged from objects at the edge of our solar system to the remotest galaxies.

2.3 The matter cycle and stellar evolution

The identification of a number of these heavenly bodies is one of the great success stories of astrophysics in the 1990s. With the help of the wide-angle X-ray camera on the Italo-Dutch satellite BeppoSAX, it was possible in some cases to localise the X-ray emissions associated with the gamma ray bursts. Shortly after each occurrence the BeppoSAX imaging X-ray telescopes were pointed in the direction of the bursts and a rapidly fading afterglow discovered, which could be localised exactly enough to identify an optical counterpart. By quickly carrying out follow-up observations in all wavelengths from the X-ray to the radio range, it has been possible to localise several gamma ray bursts in recent years. They are located in very remote galaxies. When they flare up they are the most luminous objects in the universe.

Decisive breakthrough

At present it is not clear whether these objects emit the observed radiation uniformly in all directions or only in one, or two opposed, tightly defined regions of space. It is vital that this is decided in order to calculate the emitted energy. Whatever happens, gamma ray bursts exceed supernovae in terms of luminosity. The cause of these phenomena is not clear. Scenarios are favoured in which an extremely massive star explodes in a so-called hypernova and the remainder collapses to form a black hole. However, it also appears possible that two colliding and merging neutron stars emit these enormous amounts of energy (Figure 2.48). Beside further observational data, more numerical simulations will also be necessary to shed light on this question.

Hypernova or merging neutron stars

The number of identified gamma ray bursts will increase considerably over the next decade. If they really do follow the final phase in the life of massive stars, they may be a further indicator of stellar formation in the early universe and of galactic evolution. It is therefore vital that their gamma emissions can be detected even at the most remote distances and that gas and dust in the foreground do not act to attenuate the emissions. It may also be possible to measure cosmological parameters better on a large sample of the identified gamma ray bursts displaying redshifts than with the indicators previously used. Robotic telescopes, which will be notified by gamma ray telescopes in space, will play an important role in such investigations.

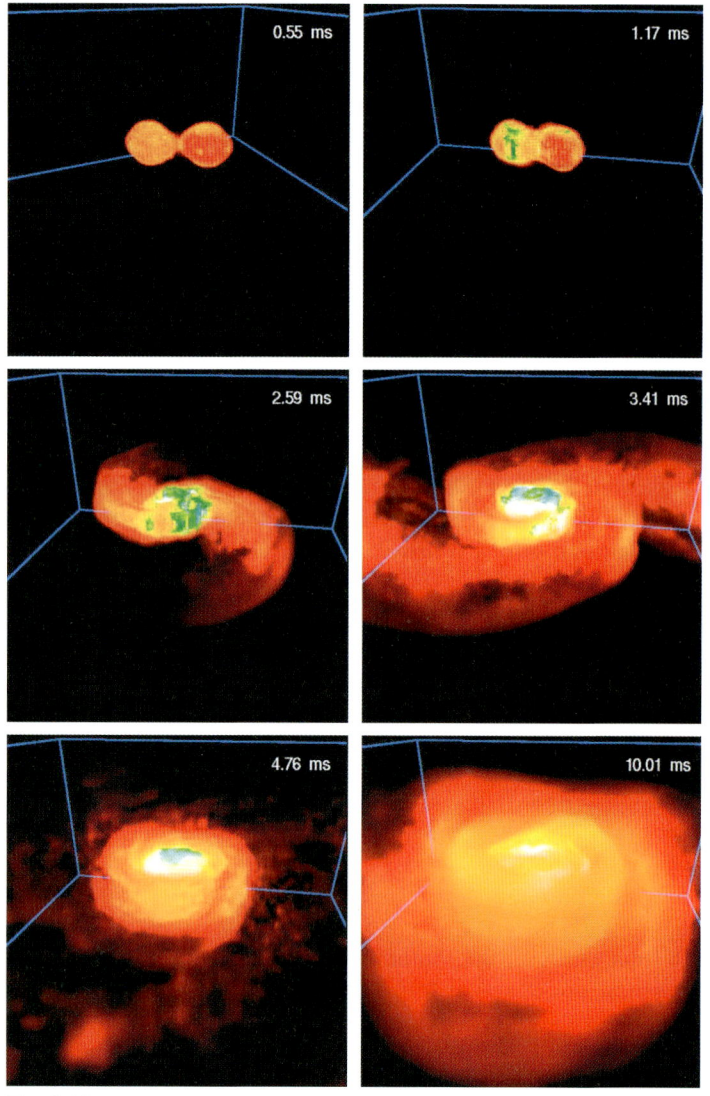

Fig. 2.48: Numerical simulation of the behaviour of two merging neutron stars (MPA)

2.4 Stellar and planetary formation: protostars, circumstellar disks and extrasolar planetary systems

Scientific advances

- Determination of the temporal evolution of the stellar production rate in the universe.
- Determination of the structure of regions of stellar genesis.
- Discovery of protostars and prestellar cloud cores.
- Discovery of circumstellar gas-dust disks and protostellar jets, numerical modelling of jet dynamics.
- Clarification of binary star statistics in various regions of stellar genesis.
- Discovery of very young stellar clusters and study of their stellar mass distribution.
- Discovery of extrasolar gas planets and the first extrasolar planetary systems.

Purpose and aims

- Clarification of the formation of massive stars and the cause of stellar genesis outbreaks.
- Deciphering the role of magnetic fields during stellar genesis.
- Proof of feedback processes during stellar genesis.
- Understanding stellar genesis in the early universe and the formation of globular clusters.
- Mass and age calibration for extremely young stars.
- Understanding the formation and evolution of planetary systems.
- Direct observation and spectroscopy of extrasolar planets and the search for biological activity signatures.

2.4.1 Star formation as a fundamental cosmological process

Star formation represents a key process for understanding the structure and evolution of galaxies and planetary systems. Because of their cosmological and cosmogenic importance, their investigation forms a focal point of worldwide interest in astronomical research.

In the past, advances in our understanding of stellar genesis were especially linked to breakthroughs in the development of astronomical observation techniques in infrared, submillimetre and radio astronomy. This applies particularly to the development of sensitive array receivers for the near and thermal infrared, and bolometer systems and linear receivers for wavelengths in the submillimetre range, and the highly successful use of the infrared observatory ISO. Using these observation options, it has recently been possible to directly observe the very early phases of stellar genesis, which occur in the interiors of very cold dust clouds. German scientists assume leading international positions in various fields of stellar genesis research. In the past it has been shown that advances in clarifying the physics of stellar genesis are only possible by close cooperation between theoreticians and observers; this cooperation is very well developed in Germany. Today, using new numerical methods and powerful computer systems, we are in a position for the first time to simulate the collapse and fragmentation process of cold molecular cloud cores, the physical and chemical evolution of accretion disks and the dynamic evolution of very young stellar clusters.

However, our understanding of the fundamental process of stellar formation has considerable gaps. For example, we do not know how massive stars form – the only stellar objects whose effects can be observed in different galaxies and which substantially influence both the thermodynamic and the chemical properties of galaxies. Some of the central physical processes – for example during the formation of protostars and the evolution of circumstellar disks to planetary systems – are not understood. In many cases, empirical rules for stellar genesis do not exist.

Only a deeper understanding of the physical processes involved in the formation of stars will allow us to extrapolate from stellar formation observed locally in regions such as the Orion molecular cloud to the governing processes in other galactic neighbourhoods, such as starburst galaxies and interacting galaxies, or to galaxy formation in the early universe. For many of the important processes involved – molecular cloud formation, molecular cloud fragmentation, effectiveness of stellar genesis, initial stellar mass distribution – neither a fundamental theory nor a phenomenological understanding exists. However, some promising fundamentals for physical models do exist.

Future development of the field

In future, excellent spatial resolution in the infrared and submillimetre wavelengths will be crucial; German astronomy will be well positioned to provide this. In the near and thermal infrared, high-resolution means the use of adaptive optics and

2.4 Stellar and planetary formation

the development of interferometers. Here, the VLTI and the LBT are available. The national interferometry centre FrInGe will assume an important coordinating function. It will be SOFIA and Herschel that will allow the first steps towards higher resolutions in the far infrared wavelength range. SOFIA will be available as a long-term observation platform in this spectral range. In the submillimetre range the *Atacama Large Millimetre Array* (ALMA) will achieve resolutions comparable to optical telescopes using adaptive optics in the infrared. The JWST will play a vital role in developing the thermal infrared, which is extremely important for stellar genesis, because it will be more sensitive by several orders of magnitude than all ground-based instruments.

In recent years German groups were frequently in overall control of building cameras for the thermal infrared; the TIMMI II instrument, which was built for ESO, is but one example. In addition, there is the GAIA space astrometry mission, which will assume an important function in calibrating the evolutionary paths of young stars (Figure 2.49).

Fig. 2.49: Measurements made by the ESA cornerstone mission GAIA will influence many areas of astrophysics: the history of star formation in our Milky Way, stellar astrophysics, the structure of our galaxy, binary systems and brown dwarfs, extrasolar planets, small bodies in the solar system, the theory of relativity and the optical reference system. (Source: ESA).

Thanks to rapid advances in computers, it will be possible in coming years to completely describe stellar genesis: from the beginnings of the formation of molecular clouds to the make-up of the finished star and stellar clusters, and the evolution of protoplanetary disks, from their formation up to the establishment of planetary systems. Special efforts are required in the extremely complex field of multi-dimensional linear and continuum radiation transport, because only by this means will it be possible to adequately interpret the observational data and compare them to the results of simulations. Advances in the development of fast numerical methods for solving the three-dimensional, non-ideal magnetohydrodynamic equations will also be extremely important for a better understanding of the dynamics of the magnetised cloud gas.

In order to draw conclusions on primordial stellar genesis it is necessary to investigate the process of stellar genesis in the environs of protogalaxies displaying only minor frequencies of metals, both by observations and theoretically.

2.4.2 Formation of low-mass stars: from prestellar core to dust disk

Complexity of stellar genesis

Stellar genesis in a molecular cloud is a complex process; in order to describe it, it is necessary to investigate turbulence, magnetic fields, heating and cooling processes and gas-dust interactions, in addition to its intrinsic gravitation. External factors are also involved: these include the formation of earlier generations of stars in the neighbourhood and their influence on current stellar genesis, the differential rotation of galactic disks, the influence of spiral density waves and interactions with other galaxies.

Formation of low-mass stars

Despite this complexity, substantial progress has been made in recent years in understanding the formation of low-mass stars of approximately one solar mass. It is now possible to observe the various phases in the formation of low-mass stars (Figure 2.50), thanks to observations using radio- and millimetre-range interferometers (the IRAM interferometer on the Plateau de Bure, for example) and the European *Infrared Space Observatory* (ISO), and using sensitive receivers in the submillimetre and infrared ranges, for example the ESO's VLT. Today's characterisation of these phases, from the prestellar core via protostars to the young T Tauri star with its circumstellar disk, is primarily based on the analysis of spectral energy distribu-

2.4 Stellar and planetary formation

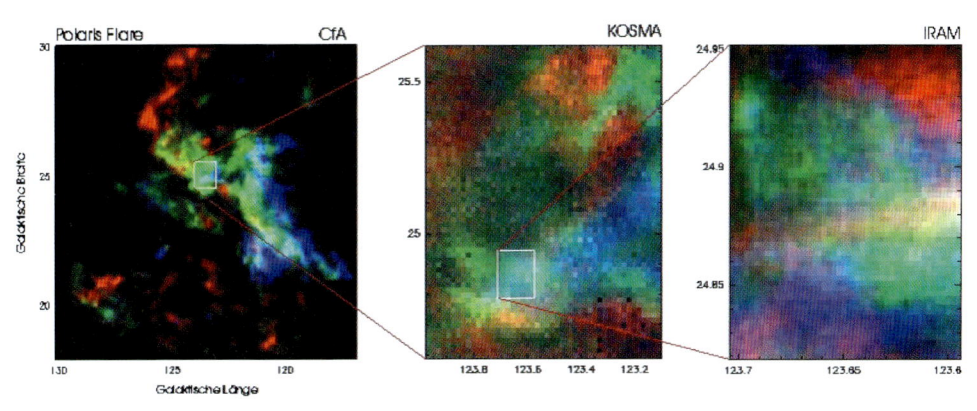

Fig. 2.50: The physics and chemistry of interstellar molecular clouds. The three images show the spatial structure of the molecular gas in a molecular cloud in our Milky Way (the "Polaris Flare"). The different colours represent the Doppler velocity of the gas. Gas that is approaching us is shown in blue, whilst gas that is moving away from us is shown in red. The left image, taken by the CfA telescope, shows the global cloud structure, whilst the centre (taken by the KOSMA telescope on the Gornergrat) and the right images (taken by the IRAM-30m telescope) resolve increasingly smaller portions of the cloud structure. The measurements emphasise the clumpy and turbulent space and velocity structure of molecular clouds, the sites of stellar genesis in our Milky Way. (University of Cologne/University of Bonn)

tions. However, it does not allow any conclusions about the respective cloud and protostar dynamics. This type of kinematic investigation is only just beginning. HST and ground-based observation – recently using adaptive optics for the first time, for example the ALFA system at Calar Alto – provide insights into the various evolutionary stages in young stellar clusters.

Investigations of young, low-mass stars are also extremely important because they provide us with data on the formation of the sun and our own planetary system, which formed together. This is very likely also the case for extrasolar planetary systems.

Today, numerical simulations can resolve the collapse of a clump of gas and the formation of a protostar with surrounding protoplanetary disk with density contrasts of more than 10 orders of magnitude. Individual phases can now be simulated, taking various physical processes into consideration, such as the turbulent fragmentation of large clouds or the action of heating and cooling processes in different phases (Figure 2.51).

In addition to astronomical observations and numerical simulations, dedicated laboratory experiments are necessary, such as those required for spectroscopic characterisation of pertinent dust particles and molecules, and for dust coagulation. This type of experiment is carried out in Germany to in-

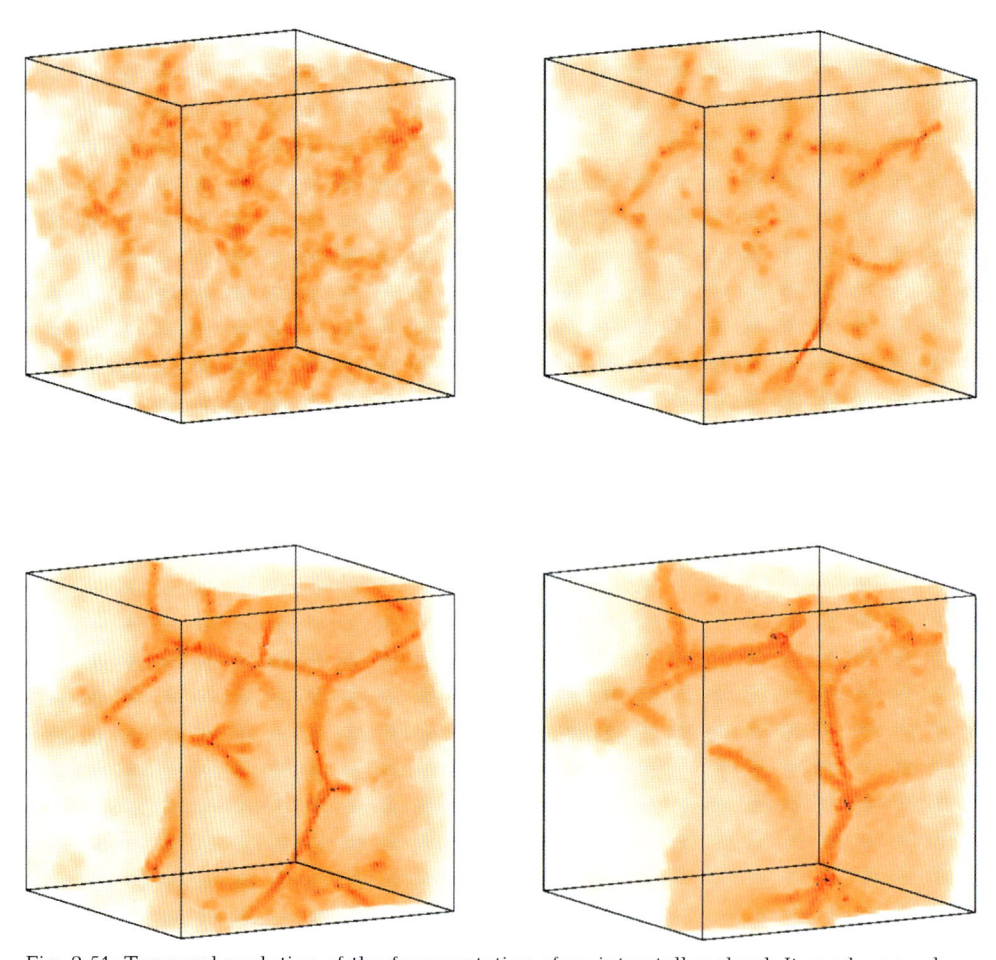

Fig. 2.51: Temporal evolution of the fragmentation of an interstellar cloud. It can be seen how the gas and dust are compacted and, in part, arranged to extended filaments; this is where stars form. (MPIA)

ternational standards and should continue to receive sufficient support.

Prestellar cores and protostars

The greatest problem when observing the initial phases of stellar genesis is that they occur in optically opaque dust clouds. In many molecular clouds there are dense, gravitationally linked objects that can only have originated in the turbulent mother cloud. If the mass of one of these cores exceeds a critical value, it collapses under the effect of its intrinsic gravitation. The gas and dust are still very cold in the initial stages of this contraction. Only recently can these objects be

2.4 Stellar and planetary formation

observed in the infrared to radio ranges. The objects are only translucent to longer-wave radiation (Figure 2.52), or the dust particles themselves emit thermal radiation. Some of the most detailed observations were carried out using the IRAM 30m telescope bolometer array and the JCMT. Characterisation of the early phases of stellar genesis took off in the 1990s thanks to the development of sensitive detectors for the submillimetre and millimetre ranges, as well with ISO. ISO was designed for the far infrared and 240 micrometre wavelengths and could therefore detect "heat radiation" from very cold dust at temperatures as low as around 10 Kelvin. This paved the way for the discovery of a whole series of prestellar compacted dust clouds. Using the submillimetre observations, it was possible to show that the mass distribution of prestellar cores correlated well with the initial mass distribution of low-mass stars.

A larger number of low-mass protostars were detected; the first examples of protobinaries are now known. It was also possible to compile spatially resolved polarisation maps using the JCMT's SCUBA bolometer system. Thermal radiation is polarised by dust particles orientated along magnetic fields, thus providing direct information on the topology of the fields (Figure 2.53). However, the search for kinematic indicators for the collapse and fragmentation processes are only just beginning. The same applies to the characterisation of the initial conditions for cloud collapse: how do cloud cores form and how are they distributed within the molecular clouds? What do their small-scale density and velocity profiles look like? How

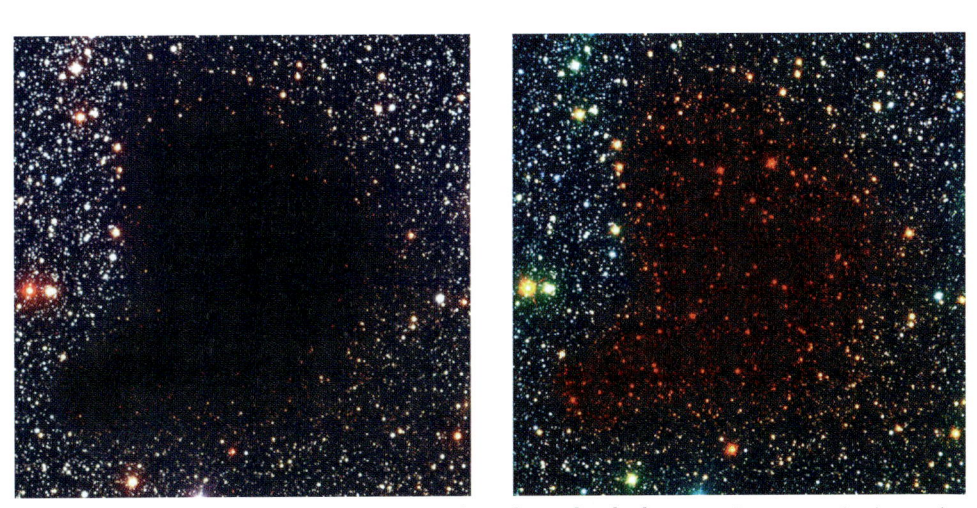

Fig. 2.52: The two images demonstrate that dust clouds become transparent at greater wavelengths: the cloud Barnard 68, left in visible light and right in the infrared, taken by the VLT. (ESO)

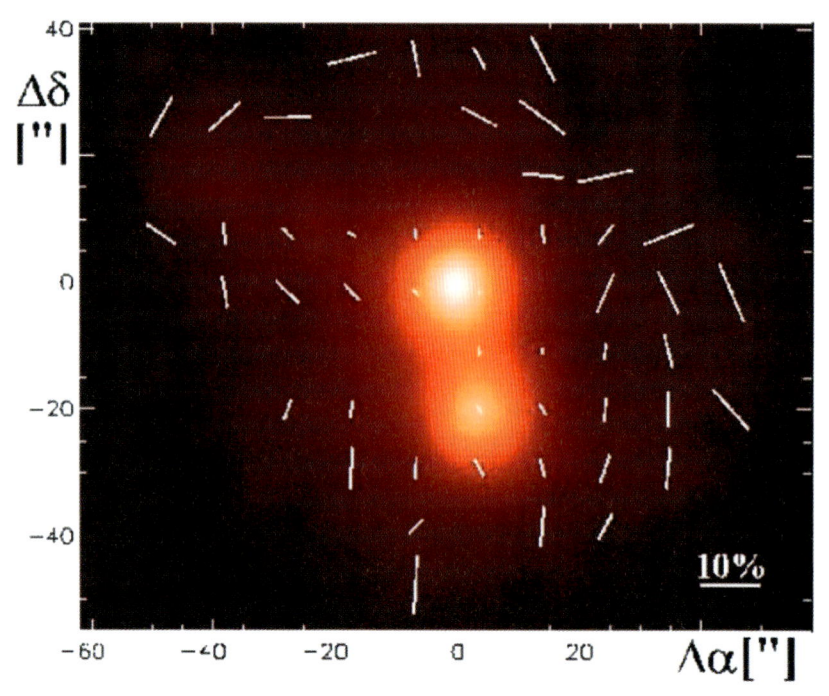

Fig. 2.53: Polarisation measurements using SC BA: intensity map with the superimposed polarisation pattern of the Bok globule DC 253-1.6 at a wavelength of 850 μm. (University of Jena/TLS)

important are magnetic fields really? Which processes lead to the simultaneous collapse of four cloud cores and thus to the formation of a stellar cluster?

Another important question is that of which physical variables determine whether stars form in single or in binary and multiple systems. Around half of all stars are binaries. Numerical simulations prove that the initial angular momentum of the cloud, its internal turbulent velocity distribution and the density distribution have a role to play here (Figure 2.54). If binary and multiple system are the rule during stellar genesis, the question of the influence of stellar multiplicity on planetary formation immediately arises. What planets form in close and in remote binary systems? What is their destiny? Are they ejected from the system by gravitational interactions? Is there, then, a population of "free" planets?

It is assumed that thousands of prestellar cores and protostars can be found in the Milky Way. They form the basis for future observations with much greater spatial and spectral resolution in the submillimetre and millimetre ranges using ALMA, and in the far infrared using Herschel. Thanks to their

2.4 Stellar and planetary formation

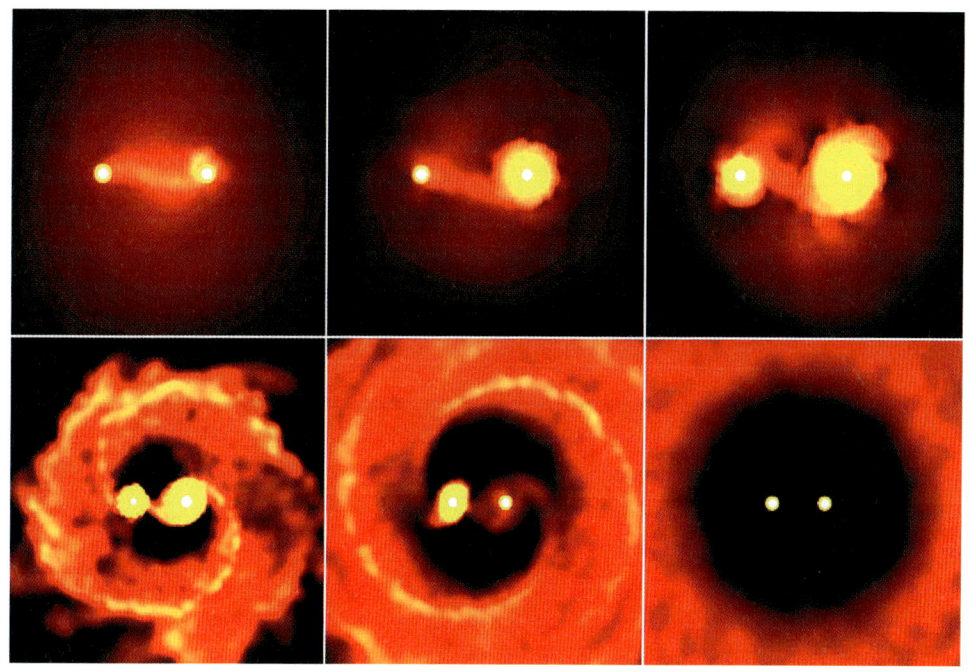

Fig. 2.54: Numerical simulations show how a multiple star system forms from a rotating cloud. A rapidly rotating disk first forms, which later fragments into several protostars. This configuration is unstable and degrades into single and binary stars. (MPIA)

participation in these projects, German researchers will have a considerable say in the investigations and be able to give answers to the questions discussed above.

Circumstellar disks

Prestellar cores rotate from their very conception, and the more they compact, the faster they rotate. Numerical simulations show that collapse eventually leads to the formation of a gas disk. In this disk, the centrifugal force and the gravitational force are in equilibrium. Matter is transported inwards in these Kepler disks, and angular momentum outwards. Planets form from the originally micrometre-sized dust particles of the disk. This is the modern paradigm, almost the same as suspected by Kant and Laplace 200 years ago. This theoretical scenario was confirmed in the 1990s, first by measurements of the energy distribution of young stars in the infrared to millimetre range, and then for the first time by direct observation using the Hubble space telescope. Direct imaging of "silhouette disks" in the Orion nebula (Figure 2.55) and the identification of disks by their scattered light in the near infrared and in optical light (Figure 2.56) were particular highlights in the search for protoplanetary disks. Meanwhile, it has been possible to image circumstellar disks in the thermal infrared and in the millimetre/submillimetre range, as well as recording velocity profiles by observing molecular lines. However, kinematic and chemical investigations of circumstellar disks are only just beginning, because of the requirement for sensitive submillimetre interferometers such as ALMA.

The transport of mass and angular momentum must be clarified with the help of three-dimensional magnetohydrodynamic simulations of circumstellar disks. The influence of

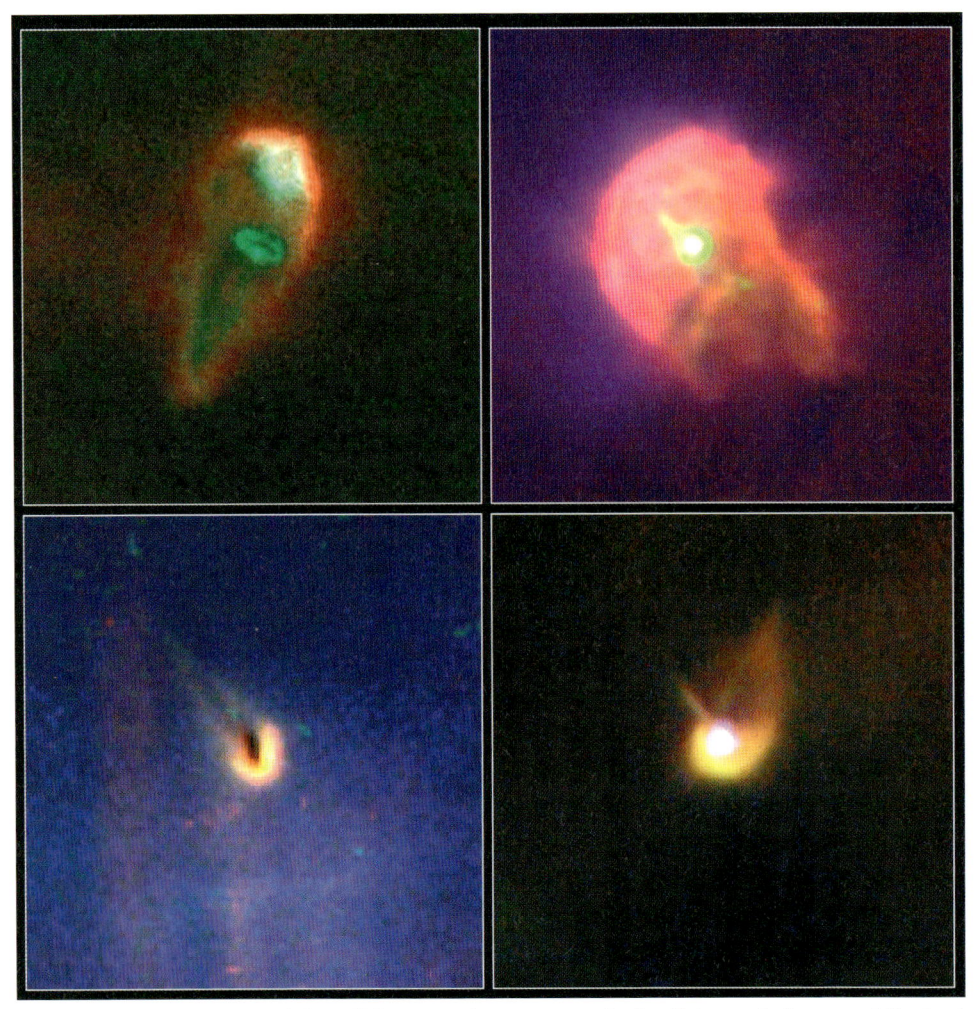

Fig. 2.55: Hubble images of dust disks around young stars in the Orion nebula, one of the best investigated stellar nurseries. (NASA/STScI/ESA/AIP)

2.4 Stellar and planetary formation

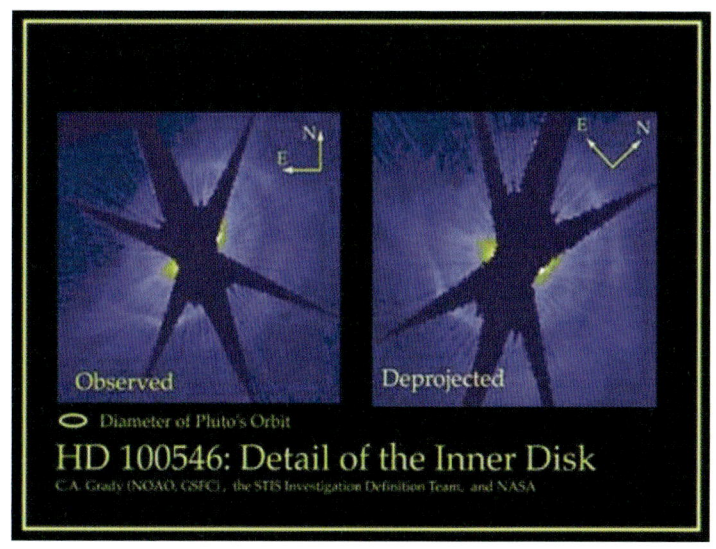

Fig. 2.56: Dust disk around the star HD 100546, measured using the STIS instrument on the Hubble space telescope. (NASA/STScI/ University of Jena)

the degree of ionisation of the disk, the development of dust and the gas-dust interactions need to be investigated. The recently discovered magnetorotational instability represents a promising mechanism for angular momentum transport in protoplanetary disks; however, there is no comprehensive proof that it solves the fundamental problem of angular momentum transport.

Jets from young stars

The discovery of tightly bundled gas jets in the mid-1980s came as something of a surprise. The supersonic jets shoot from young stars into the interstellar medium in opposite directions. The jet axis is always perpendicular to the protostellar disk. Detailed investigations of these jets (Figure 2.57) prove their connection to star formation. For example, it can be seen that the most violent emission phase occurs simultaneously with the most violent influx phase. Theoretical works lead to the assumption that jets can crucially influence disk evolution and therefore stellar and planetary evolution. Jet feedback impacting on the turbulence in the mother cloud is also probable, but not proven. By applying theoretical work and numerical simulations, it was possible to prove that the fundamental mechanism of jet acceleration and collimation is a magnetohydrodynamic process. The processes that lead from accretion to expulsion of the jet material are, however, not understood

2 The scientific issues

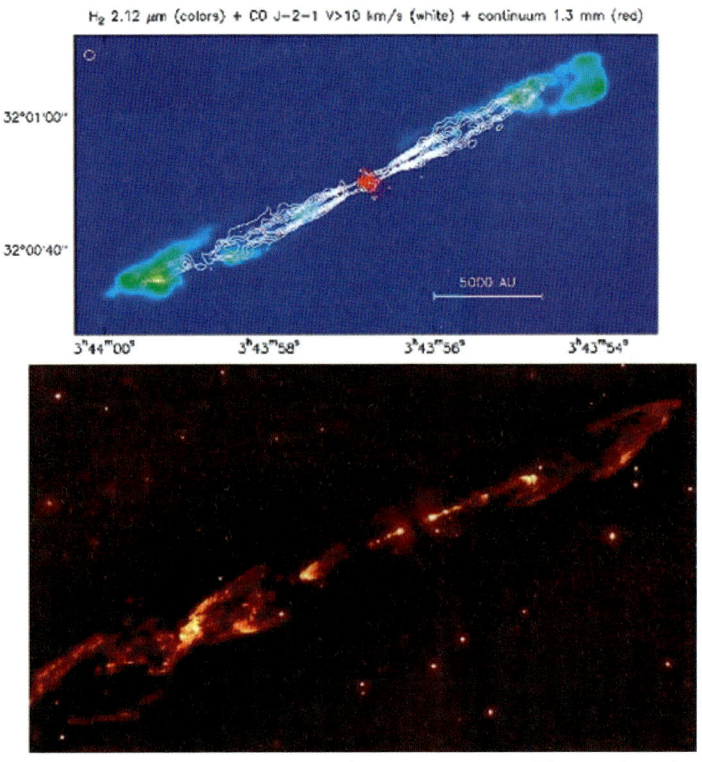

Fig. 2.57: Top: two jets, emanating from a young star. The star is at the centre, surrounded by a dense dust and gas disk; its 3 mm wavelength continuum emission is shown in red.
The CO gas (white contours) is perpendicular to the disk. Hot H_2 gas, which glows in the near infrared (green), forms where the jet impacts on the surrounding matter (AIP/IRAM). Bottom: VLT image of the symmetrical bipolar jet of protostar HH 212 (AIP).

in detail. Nor is it clear whether the jet originates at the star or in the disk. It can be assumed that every low-mass, young star (including our sun in the past) goes through a jet stage.

Statistics of young binaries

The sun is not a binary, although more than 50% of the sun-like field stars in the galactic disk are found in binary systems. In the past, a great deal of effort was invested in investigating the statistics of young binaries in nearby stellar nurseries. The development of spatial high-resolution methods, such as Speckle interferometry or, more recently, adaptive optics, proved beneficial. It was shown that the frequency of binaries amongst low-mass stars in open clusters, such as those in the Taurus-Auriga region, is considerably greater than in dense stellar clusters such as the Trapezium cluster in Orion. In the former region, the frequency of binaries surprisingly almost reaches the 100% mark; binary formation is the rule here. In contrast, the frequency of binaries in the Orion cluster is only around 50%, similar to the field stars. Among other things, this has led to the hypothesis that the field star population originates on the whole from now defunct stellar clusters – a proposal that also correlates well with other observational data, for example, that the majority of stars form in giant molecular clouds via the formation of stellar clusters. Star clusters such as Taurus-Auriga make only minor contributions to the field star population. However, they do supply the remote binaries among the field stars, which would not survive in dense stellar clusters. The reason for the difference in the frequency of binaries in open and dense stellar clusters is not yet clear. Remote binaries can easily be destroyed in dense clusters. But perhaps cluster formation itself influences the probability of binary formation.

An important task for the future will be to identify the frequency and mass distribution of the companions of massive stars and to utilise them as important indicators of the formation process. Already today it is becoming clear that massive stars possess, on average, more than one companion. These are often multiple systems consisting of a very close pair and a remote companion.

Very young binaries are ideal systems for calibrating the mass function of young stars and thus of pre-main sequence evolutionary models. Astrometric investigations using the VLTI will assume special importance for dynamic mass determination by means of analysis of the orbital movements of these objects.

2.4.3 Massive stars, stellar clusters and initial mass distribution

Although we have achieved some understanding of the formation of low-mass stars, this does not apply to massive stars. However, it is precisely these objects that we "observe" during their formation and evolution in other galaxies. The difficulties encountered in explaining the formation of massive stars are based partly on the multitude of complex feedback processes acting on the surrounding gas reservoir. Massive stars are hotter than low-mass stars and impact on their surroundings with greater radiation pressure, intense stellar winds and supersonic molecular flow. Numerical simulations need to incorporate complex radiation transport. In contrast, the nearest massive

Formation of massive stars

star formation zones are much further from us than those of low-mass stars. This fact makes interferometer investigations and observations using adaptive optics unavoidable. Finally, it should be noted that there are far less massive stars than low-mass objects in the Milky Way. We do not yet know whether massive stars are also formed by collecting matter and/or by merging young stars, or from cloud cores in the middle-mass range. In order to clarify this, it is necessary to search for the earliest phases in the formation of massive stars, for circumstellar disks and for the outflow of matter. Whether massive stars always form in clusters or can also occur isolated also requires clarification.

Improved understanding of the formation of massive stars and their mass distribution, including their upper mass limit, is also a fundamental requirement for understanding the evolution of galaxies. It is primarily the massive stars that affect the major galactic phenomena such as galactic nucleosynthesis, the production of turbulence energy in the interstellar medium, the formation of galactic winds and the formation and destruction of molecular clouds. The high formation rate of massive stars controls the entire energy balance of galaxies with high stellar genesis rates ("starburst" galaxies) and extremely high-luminosity infrared galaxies.

Stellar clusters and stellar associations

When investigating local stellar genesis, special attention must be paid to the difference between isolated star formation and that in dense stellar clusters. Infrared images of star-forming molecular clouds show that most young stars form in embedded, compact stellar clusters. In particular, the question of which processes stimulate and force stellar genesis in a given region, or which can hinder and suppress it, must be addressed. The outcome of these observations is the luminosity function and thus, via evolution analyses, the original mass and age distribution of the cluster stars.

Globular clusters are among the densest stellar systems. Their formation is still a complete mystery, but they are particularly interesting because they are among the oldest objects in the universe and are therefore relics of the first phase of stellar genesis in the cosmos. It appears that globular clusters are still forming today in "starburst" galaxies and in merging galaxy systems. In contrast to this, the densest and most massive stellar clusters in our Milky Way possess far less mass than globular clusters. They are therefore more readily torn apart in the Milky Way's tidal field and contribute to today's field star population.

2.4 Stellar and planetary formation

Beside dense, young stellar clusters, stars are also born in loose associations. With the aid of optical follow-up observations of ROSAT X-ray sources, many previously unknown young, widely distributed, low-mass stars were discovered in all nearby stellar associations (Figure 2.58).

The original mass distribution of the stars represents a decisive quantity in terms of galactic evolution, as well as that of stellar clusters. It describes the relative frequency of the stars as a function of their mass. Sensitive surveys in the optical, infrared and X-ray ranges are necessary in order to determine this, as well as spectroscopic follow-up observations. In particular, the data for the young star populations found must be as complete as possible towards the smaller masses. It has been known for a long time that the proportion of newborn stars increases with decreasing mass. However, the minimum mass to which this trend continues is unknown.

The original mass distribution of the stars

This links directly to the question of the frequency of brown dwarfs. These objects lie between stars and planets on the mass scale. They were predicted for decades but could not be proven unequivocally until the mid-1990s. The mass distribution of brown dwarfs forming the companions of sun-like

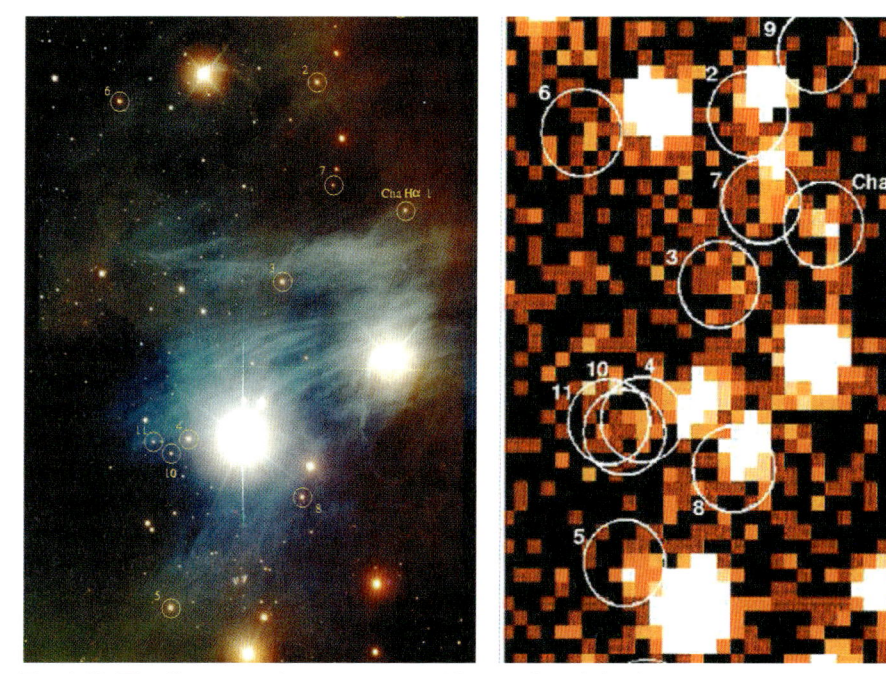

Fig. 2.58: The discovery of young stars and brown dwarfs by their X-ray activity. Left, an image in visible light taken by the VLT, right, the ROSAT image. (ESO/MPE)

stars does not appear to pass smoothly into the mass distribution of the gas giants. Rather, there appears to be a gap between the two distributions. However, these findings must still be confirmed, or refuted, in the future. Sensitive observations using the JWST will play an important role.

2.4.4 Extrasolar planets

With the direct imaging of circumstellar disks and the discovery of the first extrasolar planets and planetary systems, the investigation of forming planetary systems and the search for extrasolar planetary systems have evolved from speculation to a central research priority of modern astrophysics.

Planets such as Earth probably orbit many stars. Until recently, however, our solar system was the only one known. In 1995 it was possible to prove for the first time what astronomers had suspected for a long time: other stars are also orbited by planets. Currently, most extrasolar planets can only be identified indirectly with the help of the Doppler effect via the gravitational influence they exert on their stars. However, only objects with at least one Jupiter mass and with close orbits can be found this way. Moreover, one of these planets was recently identified by direct observation of its transit past the disk of its star, allowing the mass, radius and thus the mean density of the gas planet to be determined. The planet apparently has a density 70% that of Jupiter, but with a 40% larger radius (Figure 2.59). The discovery of the first extrasolar planets and planetary systems has given enormous drive to this field of

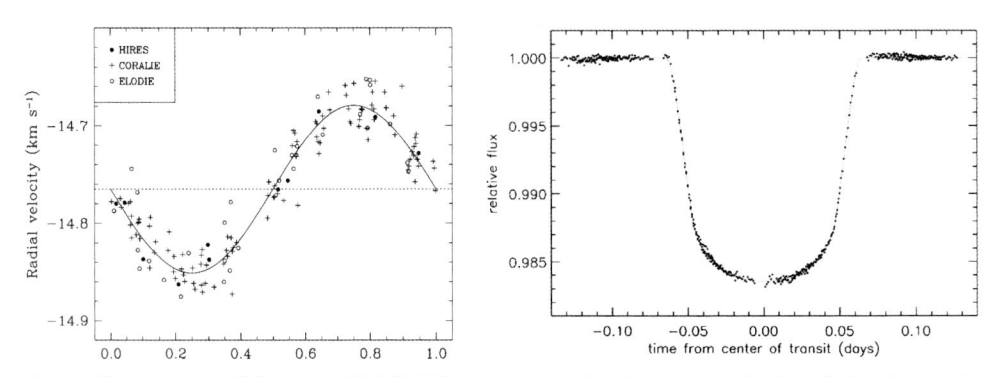

Fig. 2.59: A planet of the star HD 209458 was detected using two methods: left, by its gravitational effects on the star. The star vibrates with a speed of several tens of metres per second around its centre of gravity. On the right it can be seen how the brightness of the star altered slightly as the planet passed in front of it. It was thus possible to prove unequivocally that the planet posses 70% of the mass of Jupiter, but a 40% greater diameter. (NASA/ESA)

2.4 Stellar and planetary formation

astronomy. In particular, means of directly observing, or even investigating spectroscopically, what are currently invisible, dark stellar companions are being explored. It would then be possible to discover whether the planets posses atmospheres, their composition and whether they show signs of biological activity.

Discovery and characterisation of extrasolar planets

Up until now, only the spectroscopic Doppler method and, in the case just described, the transit method have been successful in the search for extrasolar planets. In the meantime, extrasolar planetary systems have also been discovered using the Doppler method. Its extension to cover further spectral classes will provide more statistical data on the frequency of planetary systems. The transit technique, combined with radial velocity measurements, must be extended in monitoring programmes to cover stellar clusters, in order to employ transits to discover further systems. Small robotic telescopes should be utilised for this purpose. A space mission such as Eddington would also be capable of detecting transits of Earth-like planets.

The next major objective must be to directly image planets and to investigate them spectroscopically if possible. This places enormous demands on the observation technology in order to overcome the extreme brightness contrast between the star and the planet. With the help of dedicated, high-contrast, adaptive optic systems on the VLT telescopes and on the LBT, coupled with new coronographic methods, a breakthrough in giant planets may be achieved; this will certainly be achieved using the thermal infrared camera on the JWST. Further opportunities are possible using the planned Nulling interferometers on the VLT and on the LBT. The search for the signatures of planetary atmospheres and clarification of the differences to brown dwarfs are priority research objectives. With space-borne interferometers such as ESA's future Darwin project, it will be possible to directly image and spectroscopically recognize Earth-like planets (Figure 2.60). But the so-called micro-gravitational lens effect will probably provide the first indirect proof of Earth-like planets. Numerical analyses must investigate the energy exchange and gas circulation between the hot hemisphere facing the star and the cool night hemisphere, generally facing the observer, in order to predict the visibility of the planet.

A sample of suitable young stars, as close as possible to our sun, is required in order to observe young extrasolar planets. For example, young stars that are closer to us than all previously known star formation regions have been found. They occur partly isolated from molecular clouds. These very

2 The scientific issues

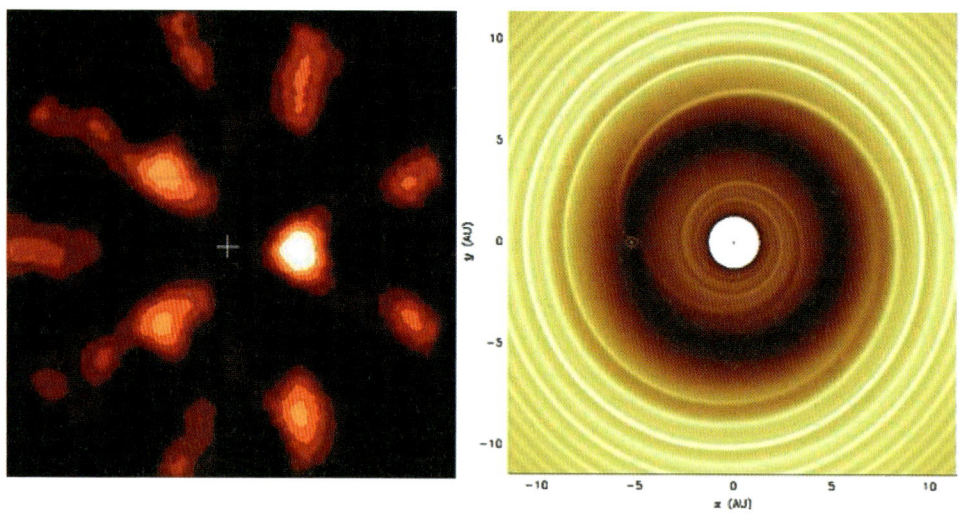

Fig. 2.60: Investigation of the formation of protostellar disks and planets. The image on the left shows a simulation of an observation of an Earth-like planet around a sun-like star using the DARWIN space interferometer. The central star is at the position of the cross and is practically completely hidden by use of so-called "Nulling interferometry", although it is 10 million times brighter than the orbiting planet (bright spot). The weaker intensity maxima are artefacts of the quite simple reconstruction methods used in this simulation. The image on the right shows a numerical simulation of a circumstellar disk, in which a dust-/gas-free zone forms in the disk due to the presence of a planet. These gaps will be visible to VLTI/LBTI, and the gas and dust distribution and dynamics in the outer zone will be resolved by ALMA. (ESA/University of Jena/MPIA)

close and young stars are especially well suited for direct imaging of circumstellar disks and possible forming planets, and for detailed investigations. This is because young planets are brighter than more evolved objects.

Formation of extrasolar planets

The formation of planets from the micrometre-sized dust particles and gas of the disk is a complex process, and one that is still not sufficiently understood today. We do not know, for example, whether gravitational instabilities in the early phase of disk evolution can lead to the formation of giant planets. Another scenario assumes that the dust particles grow to kilometre-sized planetesimals and that these then form planets by gravitational interaction and gas accumulation. The dynamic interaction of the dust particles with the gas in the initial growth phase is especially important, so the type of gas flow must be precisely known. Experimental work – including under microgravity conditions – has shown that small particles collide, adhere and grow under the conditions typical in circumstellar disks. However, if the particles reach the metre range in size, the collision velocities are so great that fragmentation must occur and the growth process should actually cease. In this case, planets would not form. Whether the fragments return to the "mother body" by aerodynamic friction or whether a gravitational instability helps here too, is currently unknown. German groups lead theoretical and experimental investigations of these processes; an experiment is currently being prepared for the International Space Station.

The discovery of planets with several Jovian masses and very close orbits around the central star, as well as extrasolar planets with highly eccentric orbits, throws up a multitude of questions concerning the formation of planets and their interactions with the circumstellar disk and with each other. The factors influencing the upper limit of the mass of forming planets must be clarified: we need to understand the conditions under which a young planet can sweep a gap in the disk when orbiting its star and how far it is possible for the planet to accumulate matter through this gap. What processes are prevalent in the Roche zone around the planet, how do circumplanetary disks form and how does the planet accumulate mass? How do planetary satellites form? These problems can currently only be approached using numerical simulations. It will be possible to detect the gaps in the disks using ALMA and the JWST (Figure 2.61). A related problem is posed by the fact that many giant planets have been found that are located very close to their star. Today's theories tell us that they cannot have formed there. It is therefore assumed that they wandered inwards during their formation as a result of interactions between the disk and the planet. But what stopped them? How often are protoplanets swallowed up by the central star?

Finally, the highly eccentric orbits of many extrasolar planets are a mystery. Do they occur due to interactions between the planets and disks, or are the interactions of the planets themselves more important? How stable are planetary systems? Why do the planets in our solar system move in practically circular orbits? All these questions will be answered during the next decade thanks to focused simulations and observations.

An answer to the fundamental question of whether conditions favourable to the development of life are the rule or the exception in other planetary systems can be expected from an understanding of stellar and planetary formation. Another important question is that of whether the astonishingly complex organic molecules, which form in interstellar matter, can reach the surfaces of newly formed planets, for example in the shape of icy cometary materials. With this question, astrophys-

Life on other planets

2 The scientific issues

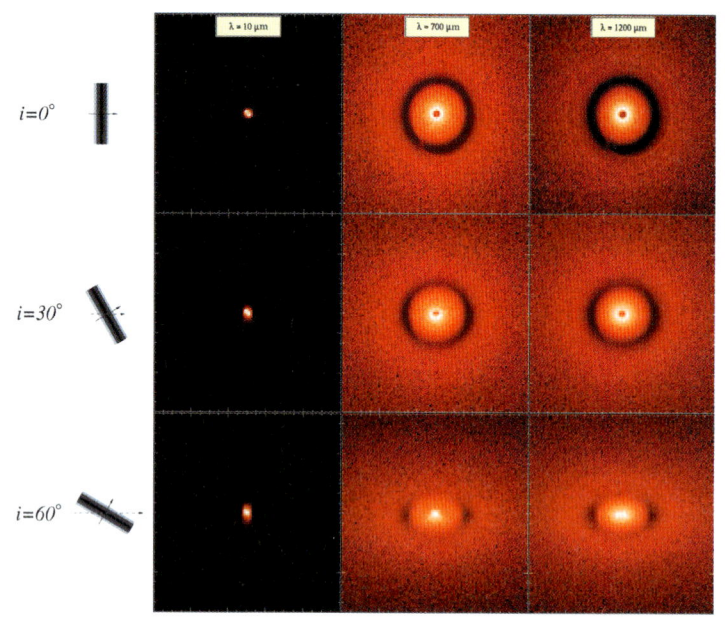

Fig. 2.61: Three-dimensional simulation of the spatially resolved emission of a protoplanetary disk with a gap, for various wavelengths (from left to right) and various angles relative to the disk plane (from top to bottom). The observation of such gaps, for example using ALMA or the JWST, would provide a direct indication of the presence of a massive planet. (University of Jena/TLS)

ics today has reached a point at which it can, for the first time, investigate the possibility of life on other planets based on reliable observational data.

Planets may live dangerously even at birth. If most low-mass stars with circumstellar disks are born in young stellar clusters, it is possible that the disks be destroyed by the UV radiation from the massive, hot stars and that no planets then form. The frequency of planetary formation is therefore by no means certain. Moreover, Jovian planets can destroy themselves by wandering inwards; they then fall into the central star and drag the inner planets in with them. Is this scenario realistic? Numerical simulations must help to illuminate this central question.

3 The next fifteen years: observatories and instruments

3.1 Access to telescopes and involvement in large international projects

We would now like to present the projects that will shape astronomy in the coming two decades. The majority are international efforts, but they also include national developments that will allow researchers in Germany to assume adequate roles in the future.

During the last decade it has become increasingly obvious that to be truly successful, the exploration of the universe and its multitude of objects and physical processes requires the use of multi-wavelength astronomy (see Figure 1.11).

Multispectral research

German astronomers are currently very well positioned in terms of the use of first-class telescopes and experiments (Table 3.1). This makes German astronomy competitive worldwide, and leader in a number of fields (see Chapter 4). The aim for the future is to not only stay apace with the rapid advances in these fields, but, as in the past, to assume and consolidate the leading positions. Past experience has shown that the international position generated by this work generally exerts a positive influence on astronomy in Germany as a whole.

3 The next fifteen years: observatories and instruments

Table 3.1: Observatories, telescopes and experiments operating during the last decade

Observatory	Field	Ground/ Space	Start	End	Funding provider	German share
Geo600	Gravitation waves	G	2002		D	100 %
Effelsberg	Radio	G	1972		D	100 %
IRAM	Millimetre	G	1979		D/F/Sp	47 %
ISO	IR	S	1995	1998	ESA	25 %
La Silla	Optical/NIR	G	1969		ESO	20 %
Calar Alto	Optical/NIR	G	1973		D/Sp	90 %
VLT Paranal	Optical/NIR	G	1998		ESO	20 %
SDSS	Optical	G	1998		USA/NASA/D	5 %
HET	Optical	G	1999		USA/D	9 %
Tenerife	Optical (solar)	G	1985		D/Sp	75 %
SOHO	UV/optical	S	1995		ESA/NASA	20 %
Hubble Space Telescope	UV/optical/NIR	S	1990		NASA/ESA	4 %
ROSAT	X-ray	S	1990	1999	D/NASA/UK	60 %
Chandra	X-ray	S	1990		NASA/NL/D	2 %
XMM-Newton	X-ray	S	1999		ESA	25 %
Compton GRO	Gamma	S	1991	2000	NASA/ESA/D	25 %
Integral	Gamma	S	2002		ESA/Russ.	20 %
HEGRA	UHE-Gamma	G	1987		D/Sp/Armenia	80 %
GNO	Neutrinos	G	1990	1997	D/I/F/Poland/USA	50 %
AMANDA	Neutrinos	G	1996		USA/D/int.	15 %
CRESST/GENIUS	Dark matter	G	1999		D/UK/I+Russ./USA	80 %

Great international efforts in the next ten years

Many of the aims of modern astrophysics can only be achieved by immense multinational efforts. This has been the case for a long time for space-based research, but also increasingly applies to ground-based activities. This trend will increase further over the next ten years, and in some fields the establishment of "global" telescopes will be required. These are such large, complex and expensive telescopes that they can only be accomplished by cooperation between research communities. The prime objective for Germany now must be to either acquire or maintain visible participation in the leading facilities and projects.

Here, the European Space Agency (ESA) and the European Southern Observatory (ESO) are of central importance for German astrophysical research, because they are the funding providers for most large space- and ground-based projects. Germany is involved as a full and equal member. In terms of maintaining international competitiveness, it is extremely

3.1 Access to telescopes and involvement in large international projects

important that German research groups position themselves well in this international environment and can acquire critical participation in order to optimise the scientific utilisation of the project results.

The most important new initiatives in the category of large international projects are discussed below. In particular, those projects that can be implemented, or at least be developed to maturity, in the next decade have been included. Of no less priority are those projects that have already reached the scientific data collection phase or will do so very soon. First and foremost here are ESO's Very Large Telescope (VLT), ESA's space observatories XMM-Newton (X-ray Multi Mirror Mission), SOHO (Solar and Heliospheric Observatory) and INTEGRAL (International Gamma Ray Laboratory), and participation in NASA's Gamma Ray Large Area Space Telescope (GLAST). They are summarised in Table 3.2.

Very Large Telescope (VLT)

ESO's VLT is the most important instrument in terms of German ground-based optical and infrared astronomy and will remain so during the next decade (Figure 3.1). In all, four 8-metre telescopes are available. Together with their excellent instrumentation, they offer improvements by factors of 3 to 50 in sensitivity, imaging speed and spatial/spectral resolution compared to existing telescopes. Second generation VLT instruments will allow further expansion of the VLT in the com-

Fig. 3.1: ESO's Very Large Telescope on the Cerro Paranal in Chile. The four 8-metre telescopes, now all operational, can be seen, together with (schematically) three of the 1.8m auxiliary telescopes, still being developed for the VLTI. Also schematically indicated are the beam combination paths for the VLTI, which are combined in a central laboratory. (ESO)

3 The next fifteen years: observatories and instruments

Table 3.2: Projects for the coming decade

Observatory	Field	Ground/Space	Start	Funding provider	German share
LISA	Gravitation waves	S	2011	NASA/ESA	13 %
SKA*	Radio	G	>2010	US/international	
Planck	Microwaves	S	2007	ESA	25 %
APEX	Millimetre	G	2004	D/ESO/Sweden	60 %
ALMA	Millimetre	G	2011	USA/Europe	10 %
Herschel	IR	S	2007	ESA	25 %
SOFIA	IR	Airborne	2004	USA/D	20 %
DARWIN/TPF*	IR	S	>2012	NASA/ESA	13 %
JWST	NIR/IR	S	2010	NASA/ESA	4 %
PRIME*	NIR	S		USA/D	10 %
LBT	Optical/NIR	G	2004	USA/I/D	25 %
STELLA	Optical	G	2003	D/Sp	80 %
MONET	Optical	G	2003	D/USA/S. Africa	80 %
SALT	Optical	G	2005	S. Africa/USA/D	5 %
Eddington	Optical	S	>2007	ESA	25 %
OWL*	Optical	G	>2012	ESO	20 %
GAIA	Optical (astrometry)	S	<2012	ESA	25 %
Sunrise*	Optical (solar)	Balloon	2004	D/US/Sp	50 %
Gregor	Optical (solar)	G	2005	D/Sp	75 %
Solar Orbiter	Optical (solar)	S	2011	ESA	25 %
WSO/UV*	UV	S	>2007	Russ./D/int.	10 %
ROSITA*	X-ray	S/ISS	2008	D/ESA	80 %
XEUS*	X-ray	S/ISS	>2012	ESA/J	20 %
MEGA*	Gamma	Balloon	2003	D/Sp/I	50 %
GLAST	Gamma	S	2006	NASA/F/D/I/J	2 %
H.E.S.S.	UHE-Gamma	G	2002	D/F/UK/Namibia	70 %
MAGIC	UHE-Gamma	G	2003	D/Sp/int.	60 %
BOREXINO/LENS	Neutrinos	G	2003	I/D/int.	10 %
ICECUBE	Neutrinos	G	2008	USA/D/int.	15 %
Pierre Auger	Cosmic rays	G	2003	USA/D/int.	20 %

A * indicates that a project is in the planning phase, but still requires approval or funding.

ing years. Thanks to the interferometric coupling of the VLT telescopes with additional telescopes, infrared observations of brighter objects with spatial resolutions of a few milliarc seconds will be possible in the coming years, an improvement of one or two orders of magnitude compared to what is possible today.

For the first time, this will endow European astronomy with the world's best observatory. German university and Max

3.1 Access to telescopes and involvement in large international projects

Planck groups are critically involved in almost half of all first generation scientific instruments (cameras and spectrographs). This puts them right at the forefront in terms of interna-tional competition, made possible not least by cooperative research. The VLT is already an enormous success.

Space telescopes

The solar mission SOHO (part of the first ESA scientific programme "cornerstone") has now provided data on the structure and activity of the sun and its corona for more than five years. German groups contributed significantly to its instrumentation. SOHO represents a substantial component of the International Solar-Terrestrial Physics Program, which was initiated by the space organisations of the USA, Japan and Europe. The influence of the sun on the Earth in particular is studied throughout this research programme.

The XMM-Newton X-ray telescope (the second ESA cornerstone, see Figure 3.2) commenced operations in 2000 and is also providing valuable scientific data. The INTEGRAL gamma ray laboratory started work in the autumn of 2002. In both of these ESA missions German groups were centrally involved in the development of the mirror system, the focal instrumentation, and especially in the development of innovative energy-resolving detectors. German groups are involved with hard- and software contributions for the very important American high-energy gamma mission, GLAST.

Herschel and Planck

In the field of large space missions, the most important for the next few years are the ESA projects FIRST-Herschel (Herschel for short, see Figure 3.3), Planck and JWST. ESA's cornerstone mission Herschel and the Planck F mission will start together in 2007 on board an Ariane 5, but then operate independently. Both ESA missions are already at the hardware stage. German institutes are leading the Herschel instrumentation and are involved significantly in analysing data from Planck.

Herschel's central objectives will include deep surveys and spectroscopic analyses of very luminous, dust-rich galaxies with high stellar production rates and black holes in the early universe. They aim to address the question of how the galaxies formed and have evolved. Moreover, detailed, high-resolution spectroscopy of stellar nurseries and the interstellar medium in the Milky Way and in external galaxies, as well as bodies within the solar system, will allow the formation of stars and planetary systems to be investigated.

Herschel, equipped with a cooled 3.5 metre telescope, will observe the cold universe with previously unattainable resolution and sensitivity in the 60 µm to 600 µm wavelength range.

Fig. 3.2: The X-ray observatory XMM-Newton during final tests at the European spaceport Kourou. (ESA)

Herschel will be more than an order of magnitude more sensitive than ISO and is capable of better spatial resolution by a factor of 5.

Planck will survey the cosmic background radiation across the entire sky, the "echo of the big bang", with an angular resolution of around four arc minutes in the 0.3 mm to 1 cm wavelength range. It will thus considerably improve on existing ground- and balloon-based measurements, as well

3.1 Access to telescopes and involvement in large international projects

Fig. 3.3: Computer image of the future European space telescope Herschel (ESA) in space near the L2 libration point. The 3.5 m telescope can be seen in the upper section, below this is the cryostat with the three instruments and the superfluid helium coolant. The lower section is the satellite bus. The satellite is protected from the Earth's and the sun's heat radiation by a large shield (which also contains the solar panels) and can therefore passively cool itself to less than 100 K. (ESA)

as on the previously started measurements by NASA's WMAP satellite. Background radiation is the earliest witness to the formation of the universe. It carries abundant information on the formation of the first structures in the cosmos and the fundamental cosmological parameters that determine the evolution of the universe.

Another very important new space initiative is the successor to NASA's Hubble space telescope, the James Webb Space Telescope (JWST, formerly NGST). It will permit the first protogalaxies and black holes, which evolved in the immediate

James Webb Space Telescope (JWST)

135

aftermath of the big bang, to be discovered and observed and will revolutionise understanding of star and planetary formation. The JWST is a passively cooled 6.5 metre space telescope. In the infrared between 0.6 μm and 28 μm it will be up to 100 times more sensitive than all previous telescopes. The JWST is NASA's central new space project and is projected to start in 2010. The ESA bodies have passed a resolution committing themselves to 15% participation in the JWST and in particular to a very significant contribution to the telescope instrumentation. An instrument in the mid-infrared range (5 μm to 28 μm), and contributions to other instruments, are to be funded from the national funds of the ESA member states; a German contribution would be welcomed.

ALMA

In the ground-based instruments field the Atacama Large Millimeter Array (ALMA) will represent an extremely important, large, new initiative (Figure 3.4). Because of its unique combination of angular and spectral resolutions and its sensitivity, it will allow the distribution and dynamics of the dust and gas mas-ses in young galaxies and their role in the formation and evolution of galaxies to be studied in detail. The same applies to galactic stellar nurseries and proto-planetary disks. ALMA will consist of 64 antennas of 12 me-tres diameter each, interferometrically coupled to form base lengths of several kilometres. They will work in the submillimetre and millimetre wavelength ranges. For example, ALMA allows recently discovered sources to be observed at resolutions and sensitivities better by a factor of 10 to 30 than is possible with currently available interferometer installations. ALMA will then be capable of discovering much weaker new objects at greater distances.

ALMA is a global project involving the USA, Europe and possibly Japan. Germany is involved through its membership in ESO and IRAM. German participation in at least 10% of the costs and thus in the subsequent observation time would strengthen and develop Germany's leading position in this field. ALMA will be funded by maintaining (and slightly increasing) the current ESO budget. In addition, it may well be in German interests to pursue additional participation via MPG and IRAM in order to secure active participation in the hardware phase and, similar to the instrumentation for the VLT, to secure access to early or privileged observations.

In this sense, the MPI (Max Planck Institute) for Radio Astronomy's Atacama Pathfinder Experiment (APEX) represents a vital contribution to ALMA. From a technical perspective, this provision of an ALMA prototype telescope represents an interesting ALMA precursor for ESO and simultaneously

3.2 Safeguarding competitiveness

Fig. 3.4: Computer visualisation of the ALMA interferometer installation. It will operate from around 2010 on the 5,000 m high Chajnantor Plateau in northern Chile (top left insert) with 64 x 12 m telescopes providing high-resolution millimetre and submillimetre astronomy. (ESO)

provides the German research community with the opportunity to perform interesting research work at the exceptional Atacama site in only a few years time.

3.2 Safeguarding competitiveness

A cooperative astrophysics research system was established on the recommendations of the last *"Denkschrift Astronomie"* (Astronomy Memorandum), and has proven to be an excellent instrument for advancement. For the first time, it has put German university groups in a position to work, on their own initiative, on substantial instrumentation projects for large international installations and thus to critically improve their scientific competitiveness. The best examples in recent years are the scientific instruments FORS1 and 2, the current "work-

Cooperative research

horses" of the VLT, developed by a consortium consisting of the Landessternwarte Heidelberg and the universities of Munich and Göttingen. Further examples include the involvement of German universities in the OmegaCAM for the VLT Survey Telescope and the LUCIFER spectrograph for the LBT.

The new astroparticle physics research programme established within the cooperative research framework takes the developments in this field into account and promotes integration of this new field of research into astrophysics in general. Traditional physics represents an important bridging element in this strongly interdisciplinary field.

In the space-based observations field, the use of national and international observation platforms by German groups has been considerably strengthened thanks to the DLR share in cooperative research. Noteworthy here are primarily the ROSAT X-ray telescope, the infrared telescope ISO, the Compton Gamma Ray Observatory and the HST (see the BMBF/DLR cooperative research report). New space-based projects (for example XMM-Newton and Integral) will be integrated in the DLR cooperative research share in coordination with the Council of German Observatories.

GRID and the Global Virtual Observatory

The new, large space- and ground-based instruments provide enormous quantities of data, mainly stored in numerous archives at Max Planck Institutes, at the ESO, and at ESA or NASA cen-tres. Improvements in archiving, handling and evaluating these copious amounts of data represent an essential undertaking with regard to the efficient utilisation of the large instruments. International and multi-wavelength evaluations of this data are only possible if the archives can be processed and compared from the respective domestic workplace. This will form the prevalent research style in coming decades. It requires compatible archive structures, very fast communications, and very large, distributed computing capacity. These requirements gave rise to the EU- and USA-funded GRID concept. It is envisaged that GRID will succeed the internet, where not only the communications with other computers are transparent to the user, but also the computer-intensive processing of external records.

Astronomy is an ideal field for testing the development of this technology. Since 1997, substantial sums have been applied for and granted in the USA, the United Kingdom and the EU for preparatory studies for just such a "virtual observatory". The concept of a "Global Virtual Observatory" has crystallised over time; its preparation will be coordinated by ESO.

The future development of the infrastructure necessary to effectively utilise this decisive instrument for interpretative astrophysics and for comparisons with theoretical simula-tions in all German institutes will be vital to keep German research competitive. This will be facilitated in the course of the "German Astrophysical Virtual Observatory" (GAVO) project.

3.3 Strengthening national innovation and initiative

National initiatives and strong participation in binational and smaller, multi-national programmes should be viewed as complementary to participation in large European and global projects. Such programmes are not only of decisive importance to a clearly identifiable German role in astronomy, they also allow particularly fast realisation of targeted experimental developments and scientific programmes. These in turn contribute significantly to reinforcing the innovation and the ability of taking the initiative of German astrophysics.

The most prominent example of such an initiative is the Large Binocular Telescope (LBT) on Mt. Graham in Arizona. This is a type of double-telescope in which two 8.4 metre mirrors are mounted on a single platform and their beams coherently combined (Figure 3.5). Due to the compact design, an unrivalled field of view is created in the infrared range, with a diffraction-limited angular resolution corresponding to that of a 22 metre telescope. The LBT will be the most powerful telescope in the northern hemisphere.

Large Binocular Telescope (LBT)

For example, in combination with adaptive optics it can observe remote galaxies in the young universe at ten times better resolution than the Hubble space telescope.

German groups have a 25% share in the LBT and already contribute significantly to the instrumentation of the telescope and its adaptive optics. By continuing with these instrumentation developments on the LBT, German astronomers can play a vital global role en-route to the next generation of large telescopes.

Another important element of German far-infrared and sub-millimetre research is the Stratospheric Observatory for Infrared Astronomy, currently under construction (SOFIA, Figure 3.6). This is a binational project involving NASA and the

SOFIA

3 The next fifteen years: observatories and instruments

DLR. It comprises a 2.7 m telescope that will fly in a specially modified Boeing 747 jumbo jet. At altitudes between 12 and 14 kilometres the atmosphere is transparent to infrared radiation; this instrument will therefore cover a wide range of wavelengths from 0.3 µm to 1,600 µm. Ten scientific instruments will be initially available, two being led by German teams.

Between the 30 µm and 300 µm wavelengths, SOFIA will exceed the spatial resolution achieved by ISO and its precursor, the Kuiper Airborne Observatory (KAO) aircraft telescope, by a factor of more than 3 and will achieve an increase in high-resolution spectroscopy sensitivity of more than an order of magnitude. SOFIA will also assume important duties in training young experimental physicists and in public relations work. It is currently planned for SOFIA to remain in service for

Fig. 3.5: The Large Binocular Telescope (LBT). Top left: LBT schematic. The two 8.4 m main mirrors are recognisable, as well as the red beam combination instrument, where the light from the two individual telescopes is combined and phased, thus creating a telescope with an effective diameter of 22 m. Bottom right: Picture of the telescope structure manufactured by Ansaldo, taken in summer 2001. (LBTB/LBTC)

3.3 Strengthening national innovation and initiative

20 years. The German share of the operations will be situated in a newly founded institute (the SOFIA Institute) and will be located at a site where active infrared astronomy is already happening. SOFIA will contribute to the fruitfulness of many fields of astrophysics and promises considerable advances, especially in the study of young galaxies and stellar nurseries.

Sloan Digital Sky Survey

A 2.5 m telescope has been operating on the whole automatically at Apache Point in New Mexico since 1998. During the next few years it will image the northern sky using five colour filters. The final catalogue produced by the Sloan Digital Sky Survey (SDSS) will consist of the positions and colours of more than a hundred million heavenly bodies, making the SDSS the most comprehensive sky survey ever. A large number of objects can already be classified based on colour. In addition, the SDSS telescope includes a spectrograph to analyse the galaxies and quasars identified in the surveyed regions. It will measure the redshifts (distances) of around one million galaxies and 100,000 quasars. The Sloan Survey will thus establish the spatial distribution of galaxies and quasars in a volume of space one hundred times greater than is currently possible. This will provide a rich hoard of data for cosmologists and for other fields of astrophysics.

Fig. 3.6: The aircraft telescope SOFIA during a test flight. The telescope will operate in the section of the aircraft marked in black. (NASA/DLR)

The project will be implemented by an international consortium of American, Japanese and German (Max Planck) institutes. The latter provide technical and financial contributions to the SDSS and are given full data access in return. In addition, a large programme of follow-up observations for interesting SDSS objects exists at the Calar Alto Observatory.

GREGOR

With regard to solar physics, the next step in angular resolution and sensitivity is represented by the GREGOR 1.5 m telescope being built on Tenerife (Figure 3.7). With the aid of multiconjugated adaptive optics it will be possible to achieve an angular resolution of around 50 milliarc seconds in the visible wavelengths, previously unattainable for ground-based telescopes. This will allow the fundamental magnetic processes acting in the solar photosphere to be studied for the first time at their intrinsic spatial scales by means of quantitative spectroscopy. This affords German solar physics the opportunity to assume a leading role, at least within Europe. GREGOR represents an important step in the development of an international solar telescope in the 4 m class, similar to that of the American-led Advanced Technology Solar Telescope (ATST). Future participation in this telescope would ensure the importance of German solar physics until well into the next decade.

Fig. 3.7: Conceptual drawing of the GREGOR solar telescope.(KIS)

3.3 Strengthening national innovation and initiative

Complementary to space-based gamma astronomy in the several tens of keV to several GeV range (INTEGRAL and GLAST), it is also possible to work on Earth's surface above around 10 GeV. Gamma astronomy involves the most energetic processes and objects in the cosmos, such as stellar explosions and pulsars, or the collapse of mass into supermassive black holes in the centres of active galaxies. The observation method used for ground-based gamma astronomy differs fundamentally from that usual for optical astronomy. Gamma rays cannot penetrate Earth's atmosphere. They collide with the atomic nuclei in the air and generate a shower of many secondary electrons and positrons. These in turn emit optical Cherenkov light, which can be detected by large, imaging telescopes.

H.E.S.S. and MAGIC

These high-energy sources are individually investigated using the HEGRA telescope system on La Palma and individual telescopes in France, the USA and Australia. However, the number of objects detected remains small. In order to facilitate the investigation of statistically relevant populations and to resolve object structures in more detail, a significant increase in sensitivity is planned for the coming generation of instruments. The limit of detection of these instruments will be displaced towards lower energies compared to today's instruments. In particular, this will allow deeper views into the universe. Multi-wavelength observations across the entire spectral range will be extremely important in the future.

Germany leads the construction of two complementary Cherenkov telescopes for next-generation gamma astronomy; H.E.S.S. (High Energy Stereoscopic System) in Namibia in the southern hemisphere (Figure 3.8) and MAGIC on La Palma. Whilst MAGIC aims at as low an energy threshold as possible in order to investigate the deep universe, the H.E.S.S. system places its emphasis on spectral and spatial surveys of extensive sources in the TeV energy range. Both instruments assume a global spearhead position and will produce their first data in 2002/2003. Further development of these telescopes can secure this position in the long term.

Some other good examples of smaller initiatives for safeguarding the ability to take the initiative of the German universities in particular include the 10 m Hobby Eberly Telescope (HET) in Texas and the Southern African Large Telescope (SALT). German groups are contributing 10% and 6% respectively for around 10 years, in part thanks to the Volkswagen Foundation. Another interesting and promising development is that of robotic telescopes. These are completely computer-control-

HET/SALT and robotic telescopes

3 The next fifteen years: observatories and instruments

Fig. 3.8: Status of the H.E.S.S. telescope installation in April 2003. Two of the telescopes, at the right of the picture, are in regular operation. Top left: details of a mirror support consisting of 380 individual mirrors. Top right: telescope camera with 960 light detectors representing pixels at the mirror focus. (MPIK).

led telescopes with specific, defined observation programmes administered by a central computer; that is, without human on-site intervention.

The possible scientific programmes range from follow-up observations of gamma-ray bursts, search and photometry of supernovae and observing the temporal variations of AGNs, to recording long-term time series' of the magnetic activity cycles of sunlike stars. At the moment, the only German installation is STELLA, consisting of two 1.2 m telescopes on the Pico del Teide on Tenerife, used for the photometry and spectroscopy of stellar activity. The availability for school educational purposes is a particular consideration of the MONET project, currently in preparation (MOnitoring NEtwork of Telescopes, one telescope each in the northern and southern hemispheres). Pupils will be in a position to develop and implement their own observation programmes via the internet and to thus be introduced to scientific research from an early stage.

3.4 Other planned space and balloon missions

MEGA

MEGA (Medium Energy Gamma Ray Astronomy) is a small gamma astronomy satellite project. MEGA will facilitate gamma ray astronomy in the 0.4 MeV to 50 MeV range and be used as a continuous total sky monitor. Compared to the precursor telescopes (COMPTEL/EGRET), MEGA will enjoy an increase in sensitivity of more than an order of magnitude. MEGA represents a technologically important stepping stone on the way to a so-called Advanced Compton Telescope, which is projected for the coming decade by both ESA and NASA experts (NAS/NRC Decadal Survey). MEGA will fill a serious sensitivity gap in the MeV range spanning two decades of multispectral observations, between the hard X-ray/low gamma range (XMM, INTEGRAL) and the high-energy range (GLAST, MAGIC and H.E.S.S. TeV observatories). Central astrophysical topics in this spectral transition zone between dominantly thermal processes and relativistic, non-thermal processes include the physics of cosmic particle acceleration (for example in the neighbourhood of compact objects and the interstellar medium, supernovae explosions and gamma ray bursts) and cosmic radioactivity as a characteristic of the creation of elements. By utilising MEGA as an all-sky monitor, unique data on the MeV range cosmic background will be gathered, beside previously unknown rapidly varying high-energy sources. In order to further develop the detector technology and test it under near-space conditions, a German-Spanish-Italian cooperative MEGA detector prototype balloon flight will take place in summer 2003.

PRIME

The proposed PRIME satellite mission ("Primordial Explorer") will survey a quarter of the sky in the 1 µm to 4 µm wavelength range using a multicolour camera on an 85 cm telescope. Because the troublesome foreground radiation can be suppressed by several orders of magnitude in a near-Earth orbit (600 km altitude), the search capacity of even a small orbital telescope exceeds that of a ground-based one. With one arc second resolution and almost one thousand times better sensitivity than the ground-based near-infrared surveys 2MASS and DENIS, PRIME will allow a breakthrough in the search for objects in the early universe with redshifts up to $z \sim 25$, as well as in the search for planet-like objects up to 150 light years distant. Unfortunately, PRIME was not selected for the NASA-SMEX programme in the summer of 2002 and will now be proposed

ROSITA

The instrument known as ROSITA (*Röntgen* Survey with an Imaging Telescope Array), an X-ray telescope on the International Space Station ISS, will carry out an all-sky survey in the broadband X-ray range from 0.5 keV to 10 keV in order to systematically identify previously concealed X-ray sources. In contrast to earlier X-ray surveys in the same energy range, ROSITA is around one hundred times more sensitive and has an angular resolution capacity around one hundred times better. It is hoped that ROSITA will discover more than 100,000 new X-ray sources, the majority being absorbed active galactic nuclei, but including several 10,000s of new galactic clusters, some of them at enormous cosmological distances. The ROSITA telescope will be accomplished by adopting an improved model of the ABRIXAS mirror system and by a substantially improved CCD detector in comparison to that used on XMM. The camera, specially designed for ROSITA, can be regarded as a prototype of the detector planned for use in the upcoming ESA XEUS mission. For example, it will be possible to considerably enhance scientific productivity compared to ABRIXAS. ROSITA was positively assessed in autumn 2001 by the DLR expert committee and recommended for acceptance in the ESA ISS programme in the spring of 2002 by the appropriate ESA bodies (AWG, SSAC). ESA will first carry out a Phase-A study on its accommodation on the International Space Station.

as soon as feasible for the NASA-MIDEX mission, in which the DLR is involved.

SUNRISE

The solar atmosphere is a complex, dynamic system. Its individual layers are magnetically coupled from the photosphere down to the corona. They must therefore be investigated as a single coherent system. Simultaneous measurements are required across as large a range of wavelengths as possible and with good spatial and temporal resolution. The balloon-borne 1 m SUNRISE telescope will observe the sun from a height of 40 kilometres. In particular, spectroscopy in the visible and UV light ranges will be possible to wavelengths as low as 200 nm. SUNRISE will also aid in technical and scientific preparations for future space telescopes on board the Solar Orbiter. It is being developed and built under German leadership and with American and Spanish participation. The combination of GREGOR and SUNRISE represents the requisites for solar physics research of international ranking in Germany in the coming decade. For example, using SUNRISE, German solar physicists will probably achieve a worldwide first in resolving the sun at the important scale at which the magnetic flux is bundled (SUNRISE will resolve 35 km on the sun). Combine this in particular with an extended wavelength range (infrared in the case of GREGOR, ultraviolet for SUNRISE) and breakthroughs are expected in magnetoconvection (plasma physics) research and research of the chromosphere, an important but little understood component of the solar atmosphere. SUNRISE will be implemented as a German-American cooperation project. Phase B has been approved by the DLR and has already begun; NASA has also approved the American contribution for Phase C/D, including the first balloon flight.

3.5 New initiatives: astroparticles and gravitational wave research

Two fundamentally new fields of physics research, particle and gravitational wave physics, have arrived in astrophysics in spectacular fashion.

In the 1990s it was possible for the first time to identify the principal component of solar neutrinos and thus to clearly delineate what is known as the "solar neutrino problem". The German-led GALLEX project in the Gran-Sasso massif, the Russo-American SAGE experiment and the Japanese Super-Kamiokande detector all play a decisive role here. In particular, the observations reinforced the suspicion that neutrinos possess a rest mass. These results must lead to an expansion of the established standard model of elementary particle physics, which regards neutrinos as massless. This new data has also encouraged a whole series of neutrino experiments, which will open up a new field of astrophysical observations.

Neutrinos

Development of the next generation of neutrino experiments, especially GNO, BOREXINO and LENS, which will determine the masses of the various types of neutrino, is a very high astrophysical priority. The technology for observing astrophysical neutrino sources, Cherenkov detectors submerged both in the polar ice and under water, is currently being developed with significant German participation.

Most progress has been made by the AMANDA project at the south pole, in which German researchers play a considerable role. Predictions of astrophysical neutrino fluxes are still very unreliable. However, it can be assumed that the next genera-

AMANDA and ICECUBE

tion detector, known as ICECUBE, with a volume of around one cubic kilometre, or the comparable European experiment based in the Mediterranean, ANTARES, will for the first time allow neutrino astronomy with a high discovery potential. From 2008 onwards, ICECUBE could survey the neutrino skies for around ten years.

Pierre Auger experiment

A different type of high-energy cosmic particle reaches the earth together with the electromagnetic signals from space. They are primarily atoms that are completely ionised due to their high energy; i.e. atomic nuclei that form what we call cosmic rays. Their spatial energy density is enormous and comparable to that of the interstellar and intergalactic thermal gas. Gamma astronomy shows that these particle components are widely distributed throughout the universe. At the highest energies the acceleration efficiency of the particles exceeds by far that of all man-made particle accelerators and cannot be explained by known cosmic objects and their associated dynamic processes. It is therefore possible that these particles originated in the big bang and reach us as the decay products of particles created at that time, and which cannot now be produced anywhere else in the cosmos. They are detected by the showers of secondary particles and light events generated when they impact the atmosphere. Investigations of cosmic rays up to the highest particle energies of around 10^{20} eV are currently only possible to a limited extent. But they do make it possible to analyse the processes prevalent in supernova remnants or in the environs of pulsars. The currently planned Pierre-Auger experiment, in which German physicists also participate, will also allow assertions on the existence of exotic particle types such as magnetic monopoles and strings.

Experimental search for dark matter

The majority of the matter in the universe very probably consists of non-baryonic dark matter, a previously unknown and unexplored type of matter, which only interacts very weakly with normal matter. On the one hand, the large elementary particle accelerators are used in the search for dark matter particles. On the other hand, dark matter, which is probably omnipresent, can be detected by its scatter when impacting on the nuclei of normal matter. The necessary large and highly sensitive detectors still require considerable technological development; examples are the low-temperature detectors such as those employed in the CRESST project or innovative silicon detectors as used in GENIUS. The impending generation of experiments may for the first time allow a positive identification of dark matter, with far-reaching consequences for astro-

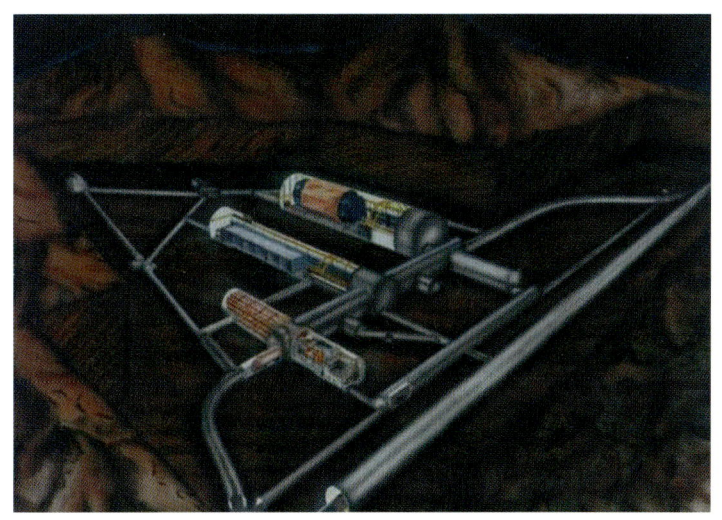

3.5 New initiatives: astroparticles and gravitational wave research

Fig. 3.9: Schematic section through the Gran-Sasso laboratory. The three laboratory caverns containing, among other things, the GALLEX/GNO, CRESST and BOREXINO experiments, can be clearly seen, as well as the vehicle tunnel. (LNGS)

physics, particle physics and cosmology. German groups are highly active in the field of particle astrophysics. The Gran-Sasso laboratory, where these experimen-ts are carried out, will also continue to provide vital functionality for particle astrophysics over the next ten years (Figure 3.9).

The most important preparatory technological steps for conducting gravitational wave astronomy have been taken. The first large, ground-based detectors, including the German GEO600 instrument in Hannover, commenced operation in 2001. The introduction of GEO600 technology into follow-up projects (LIGO-II, VIRGO-II) would very probably lead to the discovery of high-frequency gravitational waves, such as emitted by certain stars.

Gravitational waves

However, the decisive step towards detecting merging, massive black holes in galactic centres in the low-frequency range inaccessible from Earth will only be possible using the LISA space-based gravitational wave interferometer. LISA (Laser Interferometer for Space Application) is a joint NASA and ESA project. It will consist of three satellites forming the points of an imaginary triangle with sides measuring five million kilometres orbiting the sun (Figure 3.10). The distance of the satellites and thus the passage of gravitational waves can be exactly measured using laser beams. German groups are crucially involved in the design of LISA.

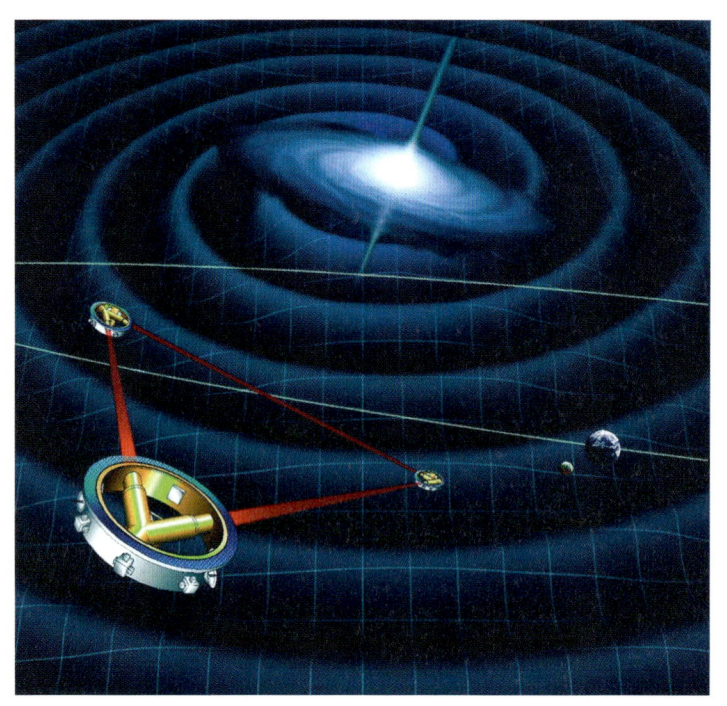

3 The next fifteen years: observatories and instruments

Fig. 3.10: Model of the LISA interferometer for measuring gravitational waves in space. (ESA)

3.6 The next decade's projects

The successful and competitive participation of German researchers in large international projects post-2010 requires timely positioning and leadership, as well as active participation in instrumentation development and the advance development of critical technologies (Figure 3.11).

Square Kilometre Array (SKA)

Radio astronomy underwent rapid technological developments as early as the 1960s and 1970s, leading to the construction of large individual telescopes (for example the Effelsberg 100 m telescope), coupled interferometers (for example VLA, MERLIN, WSRT) and to intercontinental interferometry (VLBA, EVN). Whilst one section of radio astronomy turns its attention to ever-higher frequencies, further considerable improvements are also expected in the mid- to long-term for longer wave radio astronomy. Improvements in computing power and data throughput in computer networks allow substantial improvements in the sensitivity of radio telescopes and the

3.6 The next decade's projects

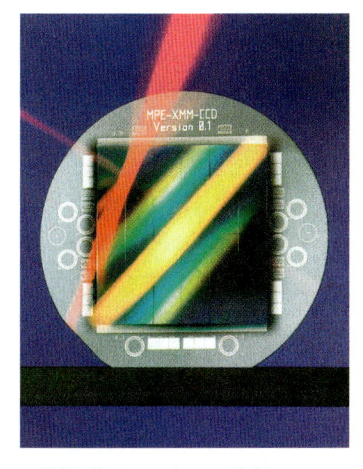

Fig. 3.11: Technological developments for the future: the imaging X-ray detector EPIC for XMM-Newton (top left), the ALFA laser for adaptive optics with the 3.5 metre telescope on Calar Alto (top right), the cryogenic CRESST detector for searching for dark matter particles (bottom left) and the MAMBO bolometer array for observations in the millimetre range with the IRAM 30 metre telescope on the Plateau de Bure (bottom right). (MPE/MPIA/MPP/MPIfR)

development of new telescope designs. By employing modern correlators and glass-fibre connections, as well as by adding further elements, the VLA and MERLIN will be more sensitive by a factor of 10 to 50, for example, and be capable of substantially improved resolutions (EVLA, e-MERLIN, e-VLBI). Individual telescopes will also become increasingly effective by using multihorn systems.

The development of innovative software telescopes will be particularly impressive. Radio waves are received by individual elements almost omnidirectionally, digitised and then correlated in a central computer. This allows several flexible antenna beams to be synthesised, troublesome radio interference to be faded out and several objects to be observed simultaneously, even retrospectively. One initial development in this direction in the near future will be LOFAR (10 MHz to 200 MHz) and later, as the technology advances, the Square Kilometre Array (SKA). More than 200 radio telescopes with an effective overall area of one square kilometre will operate here. The SKA will be around one hundred times more sensitive than currently available instruments. Together with multiple beam technology this will drastically increase efficiency, especially for survey purposes. SKA will allow investigation of the first structures in the young universe at frequencies ranging from several megahertz to gigahertz. Another application of rapid data transfer will be realtime VLBI, allowing fast reactions to explosive phenomena, for example.

Overwhelmingly Large Telescope (OWL)

These rapid technological advances will require the development of a second generation of VLT instruments in the near future in order to keep pace with global competition in large telescopes. At the same time, the studies and advance development for the generation of large telescopes following the VLT has already begun. New initiatives for telescopes with 30 to 100 metre diameter mirrors are currently being discussed.

One example is ESO's Overwhelmingly Large Telescope (OWL). Telescopes of this magnitude demand a whole series of new developments, for example for multiconjugated adaptive optics or for possible interferometric methods. Here, German groups have already provided crucial preliminary work with the development of the adaptive optics system ALFA at Calar Alto (Figure 3.11). Further development of key technologies, where German groups assume a leading role globally, is also necessary for this next generation of ground-based telescopes.

Space projects

Several high priority missions are at the ESA prequalification stage and will be studied and prepared in detail in the coming years. In the field of astrophysics these include the astrometry mission GAIA, the solar physics mission Solar Orbiter, the X-ray observatory XEUS, the interferometry mission DARWIN and the gravitational wave interferometer LISA mentioned above.

GAIA

The ESA cornerstone mission GAIA (Figure 3.12) is the next step in astrometry following Hipparcos. Currently, the mission start is planned for 2012, but may begin earlier due to the advanced stage of technological development. GAIA consists of two 1.7 m telescopes facing in opposite directions, in order to allow the positions of up to one billion stars to be measured

3.6 The next decade's projects

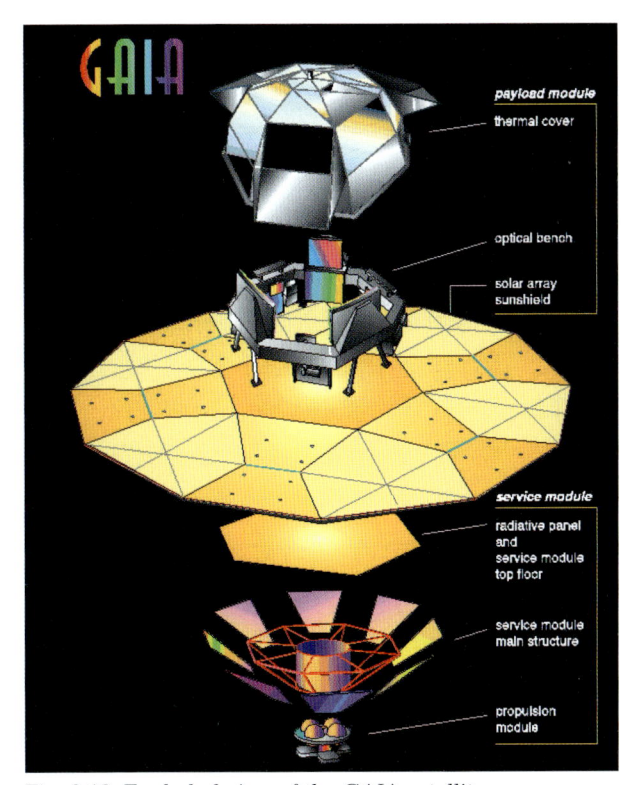

Fig. 3.12: Exploded view of the GAIA satellite.

with a precision of 10 microarc seconds. GAIA will make it possible to measure the three-dimensional distribution and the movements of stars across the entire Milky Way. This will reveal the origin and evolution of the Milky Way and the distribution of dark matter. The data precision will also allow Jovian planets around neighbouring suns to be detected, as well as enabling Einstein's relativity -theory to be tested to a precision of 10^{-6}.

Solar Orbiter

The Solar Orbiter is the next large step in solar physics after ULYSSES and SOHO. The Solar Orbiter will combine observations and in-situ measurements and approach the sun to within 0.2 of the Earth's orbital distance in order to investigate active solar regions with a resolution of only a few 10s of kilometres. The Solar Orbiter will also be capable of investigating the sun's polar regions for the first time at high resolution (Figure 3.13).

XEUS

The XEUS observatory is the next step in high-energy astrophysics; it is planned by ESA as a global cooperation (Figure

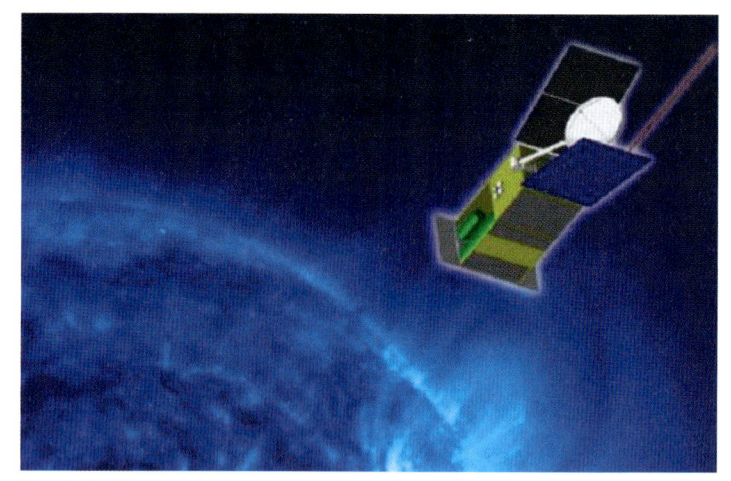

3 The next fifteen years: observatories and instruments

Fig. 3.13: Computer-generated image of the Solar Orbiter mission. (ESA)

Fig. 3.14: XEUS mission conceptualisation. (ESA)

3.14). The XEUS mirror will be transported in individual modules to the ISS and be assembled there by a robot to form a mirror system 10 metres in diameter and with a focal length of 50 metres. Such a telescope would have an angular resolution of only a few arc seconds and be around a factor of 100 more sensitive in the keV range than XMM-Newton. XEUS will discover X-ray radiation from the earliest active galaxies in the young universe and be capable of measuring directly the relativistic effects caused by the strong gravitational field of their central black holes. In addition, XEUS will provide important data for almost all fields of astrophysics. Vital technological developments are necessary in the field of fast semiconductor detectors and large, lightweight X-ray mirrors, a field in which Germany currently leads.

3.6 The next decade's projects

The ambitious DARWIN mission is aimed at searching for earthlike planets orbiting nearby stars and then investigating their chemical composition (Figure 3.15). The DARWIN mission thus goes some way to answering the age old question of the existence of other earthlike worlds and of extraterrestrial life. To this end, six 1.5 m telescopes will form a space interferometer in formation flight. It will be capable of detecting and spectrally resolving, with high spatial resolution, the weak infrared radiation of an earthlike planet very close to its star.

DARWIN

Following the projected end of FUSE in 2005, and the HST in 2010, further access to UV observations is desirable in order to maintain contact to this important spectral range on the one hand and to expertise in UV astronomy, as recently acquired by the IUE and ORFEUS missions, on the other. One planned space project in this field is the "World Space Observatory in the UV" (WSO/UV), a 1.7 metre space telescope that will observe the 115 nm to 320 nm UV range with very high spectral

Future of UV astronomy

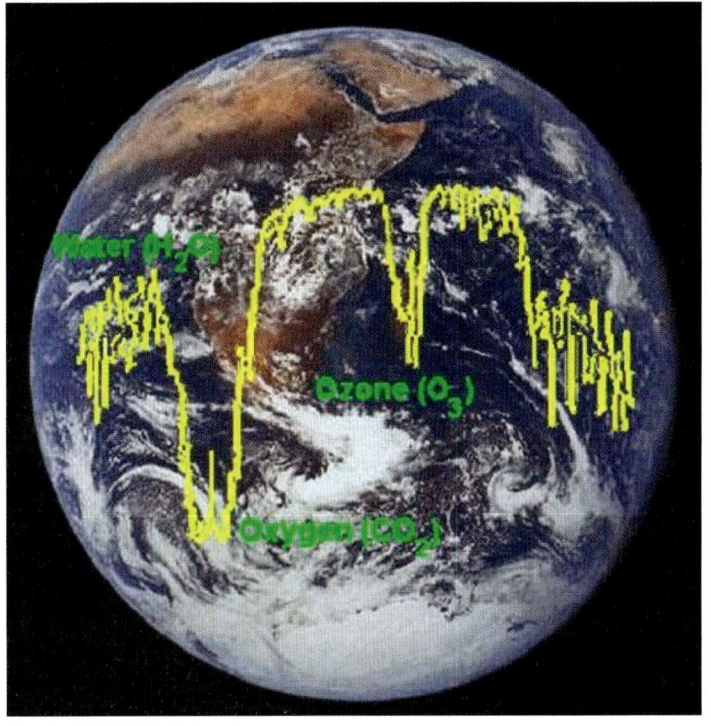

Fig. 3.15: The aim of the DARWIN mission is to discover earthlike planets around nearby stars and then to investigate their chemical composition. If these planets display earthlike conditions, oxygen-rich and ozone-rich atmospheres would be expected, with measurable signatures in the infrared spectrum. (ESA)

resolution. Compared to the Hubble space telescope WSO/UV is characterised by its doubled spectral resolution and improved spectral range coverage.

3.7 The future role of existing establishments

3.7.1 Institute for Radio Astronomy in the Millimetre Range (IRAM)

Globally leading establishments

The *Institut de Radio Astronomie Millimétrique* (IRAM) funded by the French CNRS, the Spanish IGN and the MPG operates a Volkswagen- Foundation-funded 30 m telescope on the Pico de Veleta in southern Spain (Figure 3.17, right), as well as

> **Box 3.1: Scientific yield of existing observatories**
>
> An important criterion for assessing the scientific performance of observatories is the number and rate of publications based on data obtained by the respective observatory. As an example, Figure 3.16 shows the publication rate of a number of important space observatories, as well as the VLT and the Keck telescope, as a function of the time after commencing operations. For example, the ROSAT and ISO missions, operating with strong German participation, have impregnated the field for a number of years with well in excess of a hundred publications per year, similar to the Hubble space telescope and the IUE satellite[1].
>
> A similarly important role is played by the ground-based telescopes. The publication rate of the ESO telescopes has increased to around 300 per year since the VLT opened its doors. The rates for Calar Alto, Effelsberg and IRAM are around 50 to 150 per year. This shows that past investments are paying off and have facilitated worldwide competitive research. Unfortunately, interferometer operations had to be partially suspended after two serious accidents in 1999 and can currently only be continued with restrictions.

---

[1] The publication rate is recorded with the aid of the names of the respective telescopes in the titles and abstracts in the literature accessed by the "Astrophysical Database System" (ADS), but is not free from systematic errors, particularly in the case of ground-based telescopes.

3.7 The future role of existing establishments

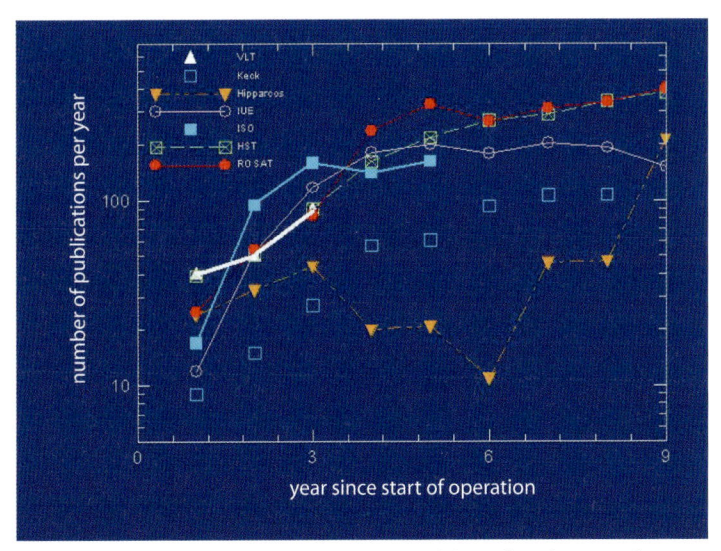

Fig. 3.16: Publication rate of several ground-based and space observatories as a function of the time after commencing operations.

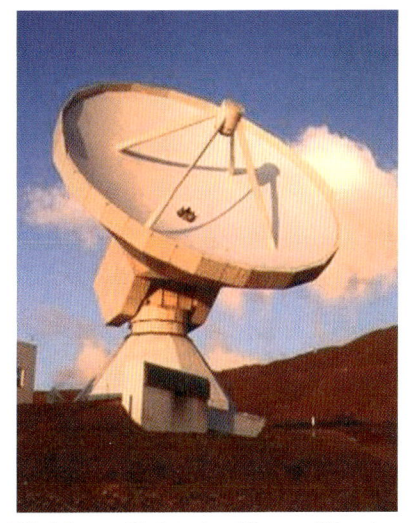

Fig. 3.17: Left: the 100 metre MPG radio telescope in Effelsberg. Right: the 30 m millimetre telescope on the Pico de Veleta in southern Spain. Both telescopes were substantially supported by Volkswagen Foundation funds. (MPIfR/IRAM)

an interferometer currently consisting of five, soon to be six, 15 m telescopes on the Plateau de Bure near Grenoble in France. Both installations are among the global leaders in this kind of telescope. The performance of these instruments has been continuously improved in recent years by development and upgrading measures, as well as by the use of more sensitive

157

Role compared to ALMA

Due to their sensitivity and geographic location, the IRAM telescopes will continue to be unique instruments for observing sources in the northern hemisphere over the coming decade. They will also continue to be extremely important for VLBI in the millimetre range. However, in the long term the interferometer installation in particular will be substituted by the new joint ALMA project. It can only be in the interests of the efficient use of funds for IRAM to participate intensely in European ALMA activities. This could be in both the hardware field (development of receiver components, integration and testing of entire receivers, participation in the development of the correlator) and in the field of software development and data evaluation. IRAM should develop a European data centre for ALMA in cooperation with other institutes.

Maintaining a high performance standard

The tasks taken on by ALMA will assume a greater proportion in IRAM activities in the future. Synergy effects will help to avoid resources for increasing the performance of existing instruments thus becoming subcritical. It should be more in German and European interests to maintain the 30 metre telescope and the interferometer on the Plateau de Bure at a high standard of performance, similar to that aimed for by the CARMA project (California Millimeter Array) in the USA. Only in this way can the requisite scientific preparation for the ALMA project be guaranteed. This includes the very important topic of training young scientists. Close cooperation with CARMA appears desirable here.

3.7.2 The Effelsberg radio telescope

The most powerful individual radio telescope in the world

After 30 years in operation the 100 m telescope in Effelsberg operated thanks to MPG funds by the Max Planck Institute for Radio Astronomy in Bonn (MPIfR) is still the most powerful individual radio telescope in the world and is Germany's only radio wavelength (from 70 cm to 0.3 cm) instrument (Figure 3.17, left). It is used by MPIfR scientists and by numerous other domestic and international institutes and universities.

Key element of the European Very Long Baseline Interferometry Network

A wide spectrum of galactic and extragalactic radio astronomy research projects is served by the telescope. With its large collecting area, the telescope will continue to play a leading role in spectroscopic observations in coming years and will be the most important element of the European Very Long Baseline Interferometry (VLBI) Network and for VLBI observations for the foreseeable future. A programme is currently running at the MPIfR with the aim of further improving the technical characteristics of the 100 metre telescope, for example in order to optimise surface properties, to employ improved and multihorn receivers and to allow observations at shorter wavelengths (below 1 cm) with greater efficiency and more dynamic observation planning. However, in the long term, the importance of the 100 metre telescope will decline with the progressing development of the Square Kilometre Array (SKA). In order to better estimate the future development and competitiveness of the telescope it is recommended that an evaluation is carried out by an expert commission in a few years from now.

3.7.3 Calar Alto observatory

For decades the Calar Alto observatory (officially the German-Spanish Astronomical Centre, DSAZ), led by the Max Planck Institute for Astronomy and primarily operated on MPG funds, has played a central role for German astronomers for observations in the optical and near-infrared, together with the European Southern Observatory on La Silla (Figure 3.18). *Central role*

In view of the numerous new large telescopes, representing a shift in MPG focus away from Calar Alto, it appears expedient to examine the necessity and future role of Calar Alto (and La Silla). Such examinations have taken place worldwide in recent years and have arrived at similar results to those in the case of Calar Alto. The deliberations of recent years in cooperation with the German (MPG, universities) and Spanish user communities and a specially instated consultative commission, show that the two large telescopes at Calar Alto will also have a valuable, if altered, role to play in the foreseeable future. This role and its future development, including in terms of the Calar Alto treaty with Spain, will be re-evaluated in 2005/2006. *Examinations*

Epochal discoveries (for example that of the extrasolar planets) have been made in recent years using small telescopes (1 m to 2 m) because the necessary long-term observations are only possible using these telescopes. Their role will now be taken over by the 2 m to 4 m class. The central tasks will include, in particular, recording long-term time series', measurements re- *Future role*

3 The next fifteen years: observatories and instruments

Fig. 3.18: Aerial view of the Calar Alto observatory in southern Spain. (MPIA)

quiring rapid reaction times and preparatory sky surveys. The use of robotic telescopes and "remote observing" will become increasingly important.

The Calar Alto telescopes also have an important role to play in innovative instrument developments, as seen in the case of the ALFA project (adaptive optics with lasers). The support of the universities should be promoted for long-term observation programmes in particular by way of them providing observers. This will also allow Calar Alto to fulfil its role in the training of young scientists.

Spanish astronomy — The rapid advancement of Spanish astronomy during the last twenty years should be taken into account by encouraging Spanish institutes and scientists to participate more actively, both scientifically and financially, in the operation and use of Calar Alto.

Synergy effects and international networking — The other European observatories (for example on La Palma) are in a similar situation to Calar Alto. It therefore appears obvious that these medium-sized telescopes should be linked to form a European network. The future instrumentation of the telescopes should be complementary and observation time be competitively exchanged between the observatories. This would allow operating costs to be driven down considerably. In order to work efficiently, the previously prevalent guest observation mode should be abandoned and observations be carried out either in remote mode or as a service offering (for example guest observers delegated for long periods).

Calar Alto cannot and need not then make all conceivable operational modes and instrumentation available, but can concentrate on its particular strengths. Despite this, the entire varied spectrum of instruments could be maintained, if necessary at different observatories, by exchanging services with other German astronomical telescopes. It is planned to implement the networking of European observatories with the financial support of the European Union.

In the foreseeable future Calar Alto will retain an important, if not decisive, role in research for a whole series of German university institutes (for example access to the northern hemisphere). With the shift of MPG's focus away from Calar Alto, greater involvement of the German university institutes (or the DFG) and the Spanish partner will be the determining factor for how and in what guise Calar Alto operations will or can be continued. This includes both the provision of personnel and operating costs.

Role of German universities

3.7.4 Solar telescopes

After the commissioning of the solar research installations at the Observatorio del Teide on Tenerife at the end of the 1980s German ground-based solar observatories were concentrated at one excellent location. The existing establishments in Locarno and Capri were closed down.

The installations on Tenerife comprise the Kiepenheuer Institute's (Freiburg) Vacuum Tower Telescope (VTT) with a 70 cm mirror, and the 45 cm Gregory Coude Telescope (GCT) belonging to the Göttingen University observatory. They are jointly operated by the Astrophysical Institute in Potsdam, the Kiepenheuer Institute, the Göttingen university observatory and, since 2000, the MPG and used by the international community of solar physicists. The French-Italian 90 cm solar telescope THEMIS now also operates at the same location. The Observatorio del Teide is now the most important location

The Vacuum Tower Telescope and the Gregory Coude Telescope

for ground-based solar research and includes several experiments for measuring solar oscillations.

The VTT can stand up to international comparisons in terms of both its size and its excellent array of post-focal instruments, and will carry this reputation on into the coming decade. One of the most important instrument developments on the VTT are the competitive adaptive optics for solar observations. The GCT, on the other hand, which is already operating in Locarno, has now passed its scientific zenith. It is currently being replaced by the new 1.5 m solar telescope GREGOR. This means that not only the infrastructure investments already made are being effectively used. The leading position of experimental German solar observations will be expanded by the planned completion of GREGOR in 2005 and the excellent cooperation with Spanish colleagues will be further reinforced.

4 Astronomical research structures

4.1 Historical development

From relatively modest beginnings after the Second World War, astronomy and astrophysics have grown from a sideline at a few German universities into an important branch of physics research. At the core of this very positive development is the strong tradition in stellar physics and theoretical astrophysics at a number of German universities. A substantial new element was added by the growth of astronomy and astrophysics in non-university establishments, in particular the Max Planck Society (Max-Planck-Gesellschaft, MPG). Today, astrophysical research in Germany can compete with the world leaders in a whole series of fields and even leads some of them. Its strong points are concentrated in small centres and research groups specialising in just a few topics.

Funding instruments

Beside the heavy involvement of the MPG, the post-war growth of DFG and BMBF (including DLR) funding instruments has played a decisive role. For example, they funded German membership in international organisations, especially the European Southern Observatory ESO and European Space Agency ESA, and made available central, national funding instruments such as collaborative research centres, priority programmes, cooperative research and national projects in extraterrestrial research. The support of the BMBF, the DLR, the MPG and a number of foundations (VW Foundation, Krupp Foundation) also made it possible to fund several large, and numerous small, projects in which Germany is either exclusively involved (for example the Effelsberg radio telescope) or as a strong leading partner (Calar-Alto observatory, Institute for Radio Astronomy in the Millimetre Range IRAM, Tenerife solar observatory, the ROSAT X-ray satellite, the Heinrich Hertz telescope).

New fields of research

The increasing interest of particle and nuclear physicists in astrophysical problems is only one of the important developments of recent years. It has led to the formation of the new research field of astroparticle physics, both in Germany and internationally. It deals with problems in which particle physics and astrophysics complement and mutually stimulate each other. It includes the neutrino experiment GALLEX and the HEGRA high-energy observatory.

Numerical astrophysics

A similar synergy effect can also be observed in the interaction of astrophysics and numerical methods: the evolution of numerical mathematics has advanced far enough that its methods and procedures can be applied to the solution of realistic astrophysical problems. This mutual interest has been promoted in the context of various collaborative research centres and DFG priority programmes.

Increase in extragalactic research and cosmology

A distinct increase in the pursuit of extragalactic and cosmological problems has been noted during the past ten years. This field was only poorly represented in German astronomy before this time. However, it can be seen from publication and impact statistics (Box 4.1), among other things, and from the increase in observation time acquired at the large observatories against international competition, e.g. at ESO or on the Hubble space telescope (see the BMBF and DLR cooperative research report), that Germany also occupies a strong international position in this field. The number of publications based on ROSAT observations is also a clear indicator of this.

Box 4.1: The impact of German astrophysical research during the last fifteen years

One important criterion for the competitiveness of a scientific community is the impact of its publication activity, measured by the number of citations by other workers active worldwide in the field. In this kind of "bibliometric" comparison it is almost impossible to eradicate a number of distorting factors, such as citation traditions in various fields and countries, and the normalisation of the number of citations in terms of the size of a community, etc. Nevertheless, the number of referenced works that are cited often has crystallized as a criterion of success that transcends field boundaries. In this context, if the more than 19,000 referenced publications with authors in German institutions listed in the ADS database (Astrophysical Data System) since 1985 are investigated, 110 publications with more than 100 citations are found for the period. These "high impact" publications are distributed across the fields given below.

Sorted according to the scientific topics covered by this memorandum:

- The universe – its origin, evolution and large-scale structure: 25
- Galaxies and massive black holes: 31
- The matter cycle: 29
- Stellar and planetary genesis and extrasolar planets: 17
- Miscellaneous: 8

Divided under observational astronomy into the various wavelength ranges and according to theory:

- Gamma-ray astronomy: 9
- X-ray astronomy: 16
- UV astronomy: 4
- Optical astronomy: 23
- Infrared, submillimetre and millimetre astronomy: 17
- Radio astronomy: 5
- Theory, including numerical simulations: 31
- Miscellaneous: 5

The division into observational fields generally reflects the technological developments of recent years in the various wavelength ranges.

4.2 The current situation

Because of the investments in institutions and personnel mentioned above, as well as in infrastructure and instruments, German astronomers and astrophysicists are today in a position to carry out cutting-edge international research in a number of fields. Beside the traditional field of stellar physics these include numerical astrophysics and especially the new observational astronomy fields, including radio, infrared, X-ray and gamma-ray astronomy.

Research centres

Garching near Munich now hosts one of the leading centres worldwide for astronomical research. This is also the case to a lesser degree for Heidelberg, Bonn and Potsdam/Berlin. The scientific significance of these sites is partially a result of the presence of one or more Max Planck Institutes with an astrophysical focus. As the uppermost body, the Council of German Observatories (Rat Deutscher Sternwarten – RDS) today has 35 member institutes. Ten have joined in the last ten years alone as a result of reunification, but also due to expansion into new fields of research (Chapter 6 contains a detailed breakdown of the relevant figures).

Significant fields of research

A few of the fields in which German astronomy assumes a recognized leading international position (publications, citations, invited lectures, acquisition of international observation time and funds, international cooperation) are listed below (see Box 4.2). In the scientific field these primarily include:

- observational and theoretical stellar astrophysics, including the sun and the final stages of stellar evolution,
- astrometry and stellar statistics,
- black holes and galactic nuclei, galactic evolution,
- numerical astrophysics,
- stellar genesis, the interstellar medium and astrochemistry,
- observation and theory of the evolution of cosmological structures,
- gravitational lenses and gravitational wave research,
- solar neutrinos.

4.2 The current situation

Box 4.2: Astronomy funding in Germany: an international comparison

Astronomy funding in Germany amounts to €226 million p.a. (see Table 5.7 for figures for 2000), i.e. €2.75 per capita. This per capita sum is somewhat lower than for our European neighbours, who spend around 2% (United Kingdom), 18% (Italy) and 29% (France) more per capita on astronomy. The figure is around 51% greater in the USA; Australia and Canada spend only 32% and 20% respectively of what Germany spends (source: Canadian astronomy memorandum). The relative uniformity among the European countries is the result of common ESO and ESA membership, among other things. In contrast, the *personnel situation* in German astronomy is much worse compared to the rest of Europe. A few comparisons will demonstrate this.

If International Astronomical Union (IAU) members are considered relative to their respective populations (or the gross national product), Germany occupies a position in the lower third (see Table 4.1). (The members of the IAU are appointed by the IAU as nominated by the respective national bodies; in Germany this is the Council of German Observatories). In the United Kingdom the number of permanently employed astronomers increased between 1994 and 2001 by 24% to 360 (if Advanced Fellows and Royal Society University Research Fellows are included the increase is 21% to 411). This is in contrast to 472 established posts in the astronomy field in Germany, of which an estimated 350 are permanently occupied. Otherwise, the numbers stagnated between 1989 and 1999 (for an approximately 35% higher population following reunification). France enjoys 720 permanently occupied astronomical posts, that is, around twice as many as in Germany, despite an approximately 28% smaller population. However, the fact that there are practically no secondary funding sources in France must also be taken into consideration in this comparison.

This weakness in the personnel situation is primarily due to the current evident under-representation of astronomy within physics in comparison to other countries. Whilst there are approximately 1,303 physics professors at German universities, the number of astronomy professorships at universities numbers 46, i.e. around 3.5% of the total. If non-academic astronomy professorships are included, this proportion increases to around 7%. In contrast, the proportion of astronomers in permanent positions within physics in the United Kingdom has increased from 21% (1994) to 29% (2001). This under-representation of astronomy within physics in Germany can also be seen in the context of some average European values: whilst barely 6% of physics dissertations in Germany originate from the astrophysics field, the relative proportion of astrophysical applications to the EU's Marie Curie Fellowship Programme (where the applicants are generally fresh doctorates) within the physics field is around 18%; this number represents a fairly reliable estimate of the proportion of astronomical doctorates within the field of physics in the EU.

In the technology and instrument development field, for example, German research leads in the following fields, which are important for future development:

- X-ray mirror and energy-resolving detectors in the X-ray and gamma-ray range,
- high-resolution infrared and optical astronomy (including spectroscopy, adaptive optics and interferometry),
- optical and infrared instruments for large telescopes,
- detectors and instruments in the far infrared, submillimetre and millimetre ranges,
- intercontinental radio interferometry,
- radiochemical methods of isotope isolation, cryogenic dark matter detectors,
- laser interferometry for measuring gravitational waves,
- stereoscopic methods using Cherenkov telescopes and air shower arrays,
- algorithms and software for astrophysical simulations.

4.3 Research institutes

Today, astrophysical research forms the main theme in the work of twenty-three university institutes, five MPG institutes and two Leibniz Societies, including the Astrophysical Institute in Potsdam (AIP), added after reunification. Three state institutes also exist, including the state observatory in Tautenburg, Thuringia. In addition, there are three non-academic institutes where astrophysics represents an important part of the research programme (also see Tables 6.1 and 6.2). A total of 524 scientists are employed at these institutes, 375 at established posts, 47 as associate professors and 50 as full professors. A further 624 doctorates and graduates are included, and 405 graduate engineers and technicians. This makes the German astronomy/ astrophysics group one of the largest in Europe in terms of numbers, although it is still in the lower third in terms of population and gross national product (see Table 4.1). The research situation at university institutes and Max Planck Institutes is described in more detail below.

University institutes   Astronomical research at universities is generally located in separate astronomy institutes and, in some cases, as groups within physics institutes. The university astronomy institutes are integrated in the respective physics faculties. Joint lectures and colloquia allow contact between physicists and astronomers, and new fields of research such as astroparticle physics and gravitational wave research intensify the cooperation.

The university institutes have a research and education undertaking that is highly fruitful in this form of coexistence. Research is strengthened enormously by graduate and doctoral student education on the one hand, and would be almost impossible to imagine without the work of students. On the other hand, the lively research activity of university lecturers is an excellent prerequisite for ensuring that current research topics are competently dealt with and that students can complete their theses in active working groups. Astrophysics is consciously perceived as an attractive subject by students, as reflected in the increasing number of physics students taking astronomy as an optional subject.

Institutional funding is the responsibility of the federal states. During recent years university institutes have been forced to accept considerable cutbacks in both budget and personnel in some areas (see Table 6.3 for example). This has led to further worsening of the existing failings in the institutional infrastructure. Most German astronomy institutes are both too small and too poorly equipped. They are lacking the most fundamental resources: travelling funds, library budget, IT funds, visitor funds. The reductions in staff posts, which have led in particular to a dramatic decrease in the scientific middle tier, occurred and continue to occur in an epoch in which astronomy is enjoying enormous worldwide advances (Chapter 2).

In terms of personnel, German astronomy now looks internationally bleak as a consequence of these reductions (see Box 4.2).

Yet another consequence of the reduction in staff at universities is the small proportion of university institutes involved in the total astronomical research in Germany: around 40% of all full professorships, almost 70% of all established scientific posts and more than 70% of the total German astronomy and astrophysics budget is situated outside of the universities (for comparison: around 70% of public expenditure for education and research goes to the universities).

The flexibility and performance of German university institutes is excessively dependent on external funding. This can be illustrated by the case documented in Table 6.2, showing that 85% of university doctorates are financed through external funds. A student's decision to work on a project is therefore often adversely influenced by the long processing

Table 4.1: IAU membership

	Number of IAU members	Population (millions)	Gross national product (trillions of US$)	IAU members per million population
Sweden	100	8.9	0.23	11.2
Switzerland	80	7.3	0.29	11.0
The Nederlands	172	15.7	0.39	11.0
France	643	58.9	1.43	10.9
Belgium	101	10.2	0.26	9.9
Denmark	52	5.3	0.18	9.8
United Kingdom	561	58.7	1.26	9.6
Greece	99	10.6	0.12	9.3
USA	2,300	276.2	8.08	8.3
Italy	437	57.3	1.15	7.6
Germany	455	82.1	2.18	5.5
Spain	218	39.6	0.56	5.5
Austria	32	8.1	0.22	4.0
Japan	471	126.5	4.09	3.7
Poland	121	38.7	0.15	3.1

SOURCE: http://www.iau.org/IAU/Organization/admdoc/adhering.html and Aktuell 2001, Harenberg Verlag

and decision times involved in conjunction with external funding applications, which can often be in excess of six months. The lack of planning security in a research group generally dependent on external funding impedes competition for the best scientists: recruitment advertisements with their required lead times are therefore difficult and rare at universities. The problem is forced even further by the falling approval rate of DFG applications following standard procedures. The necessity for flexible handling when employing doctoral students has been recognised by the MPG and taken into account by founding International Max Planck Research Schools; DFG Research Training Groups also offer the necessary flexibility within a given funding period.

The regulations at German universities make it difficult or even impossible for professors to undertake long secondments. In contrast to their European or American colleagues, leading German scientists can rarely take advantage of opportunities for limited periods of research at international establishments such as the ESO or other universities and MPIs.

4.3 Research institutes

The inadequate representation of German scientists in leading positions at the ESO (of the eight people in the ESO management none are German) and the ESA can in part be traced back to this problem.

Max Planck Institutes

The establishment of a number of Max Planck Institutes in the 1960s contributed enormously to Germany's restoration as a strong astronomy nation. The fact that the large astronomical instruments had to be rebuilt after the war and that the requisite institute sizes could not be achieved at the universities was of central importance to this development. The erection of the Calar Alto observatory and construction of the largest fully steerable radio telescope in the world, Effelsberg, are just two examples of this.

Research financing at the Max Planck Institutes is primarily funded by the Max Planck Society; they reacted to the worldwide surge in astronomy by encouraging its growth within the MPG.

A number of directors, each assigned to a separate department, work at the Max Planck Institutes. Depending on the type and size of the institute, the MPIs host technical departments and working groups for young scientists. The ratio of established scientific posts to the number of directors varies between the respective MPIs, but is generally at least seven. In addition to these institute posts, the MPIs are also increasingly acquiring external funds (for example from the EU or from cooperative research), although these still only represent a small proportion of the total funding of the MPIs.

In contrast to the universities, the MPIs can, on the whole, concentrate solely on research. The respective astronomical MPIs are located at sites hosting large universities which offer astronomy as a subject. The MPIs are linked to the universities by the fact that some of the scientists are readers or honorary professors at the universities and are involved in teaching. This provides an opportunity for graduate and doctoral students training at the MPIs. The founding of two astronomical International Max Planck Research Schools in Garching and Bonn, Germany, will further strengthen the link between the MPIs concerned, and the universities and their training activities.

Access to large astronomical devices

Experimental astronomical research necessitates observations using modern telescopes. German astronomers can access these telescopes by a number of means. The two most important are:

- Membership in ESO and ESA: access to the ESO telescopes is vital for optical astronomy. The two ESO observatories on La Silla and the Paranal host a wide range of large telescopes with modern instrumentation. The use of four 8.2 m mirrors on the Very Large Telescopes and the powerful instruments of the VLT allow German astronomers to perform absolutely top-class research. The long-term German share of the applications submitted and approved from all member countries is around 30%; i.e. considerably higher than the German share in the membership contributions. The ESA space research programme allows German astronomers access to modern space-based instruments. Observation time is awarded by application for many of the ESA missions, similar to ESO. The applications are considered by merit of their scientific quality and uniqueness. ESA missions such as Hipparcos or ISO have contributed substantially to research in Germany; with the XMM-Newton satellite, which started at the end of 1999, European astronomers are now in an excellent position in terms of X-ray astronomy. Germany also assumed a preferential position due to the extremely positive experience gained with ROSAT. The ESA share in the Hubble Space Telescope guarantees access to this telescope, which is so important for UV, optical and near-infrared astronomy. With eight accepted programmes during the last application period (cycle 11), German researchers enjoy the largest proportion after the USA.

- Telescopes with German involvement: building the Calar Alto observatory was a decisive step towards facilitating competitive optical astronomy observations. The same applies to building the Effelsberg radio telescope, involvement in IRAM and the HHT, and building and operating ROSAT. These large projects were each executed by Max Planck Society institutes, in part with considerable external funding (BMBF, DLR, Volkswagen Foundation). A significant portion of the observation time on many of these instruments is open to the entire German astronomy community upon application.

Further access to telescopes is possible as a result of the involvement of either individual institutes or groups of institutes in observatories such as the Hobby Eberly Telescope or the future LBT. However, this access is generally restricted to the institutes involved. Most large observatories award part of their observation time to observers not involved with the funding institutes: this means that German researchers can therefore

4.4 Research funding instruments

also apply. The difficulties involved with this type of access are manifold and extremely difficult to overcome in some cases.

Beside basic institutional funding by the MPG and the German federal states, and contributions to the international organisations ESO and ESA, astronomical research in Germany is also supported through other instruments. The most important of these are introduced below.

Astronomy and astrophysics cooperative research

The introduction of cooperative astronomy and astrophysics research by the BMBF as a consequence of the recommendations of the last memorandum (1987) gave astronomy a very tangible qualitative and quantitative underpinning. One of the fields in which cooperative research is carried out is "Large Device Astronomy", which supports access to large national and international ground-based telescopes. Among other things, this led to German universities being in a position to develop large instruments on their own for the first time. Some excellent examples include the state observatory (*Landessternwarte*) at Heidelberg and the university observatories in Göttingen and Munich, which developed and built the world class FORS1 and 2 instruments for the VLT. Technologically speaking, these belong to the highest class of instruments.

The second cooperative research field principally supports the use of large international space observatories. Thanks to this sponsorship those scientists who are successful in the international competition for observation time are not only given the necessary planning security, but also enjoy significantly faster and higher-quality project execution, leading to better placing among the international competition.

The rapid developments in the field of cooperative research have been taken into account by the newly introduced field of astroparticle physics; it is expected that a considerable surge in innovations will result from this research field, situated as it is between particle physics and astrophysics.

The positive effects of these examples, including in student education, are enormous, as is evident from the increase in doctorates in the astrophysics field (see Table 6.4). These developments have nevertheless proven to be of greater benefit to the large German university groups, which had already achieved a certain critical mass.

Collaborative Research Centres, Priority Programmes and DFG Resarch Training Groups

The BMBF's and DLR's cooperative research has proven to be an extraordinarily important supporting element in terms of the utilisation of large devices (especially for the universities). Further extremely important and tremendously successful instruments include the Collaborative Research Centres, Priority Programmes and DFG Research Training Groups. The importance of these instruments lies in their capacity to strengthen interdisciplinary cooperation between researchers of different fields and the coordination of long-term projects between institutes. These elements come to particular fruition in the field of astrophysics. Astrophysical research is based on the interaction of a large variety of physical aspects on the one hand. On the other, the German astronomy community is divided into numerous relatively small units and only achieves supercritical mass and synergy effects by way of these interactions.

The introduction of the highly successful DFG Priority Programme 450 "Theory of cosmic plasmas" (1987–1993), for example, stimulated by the last memorandum, has given a gigantic impulse to further astrophysical research. The DFG Priority Programme 451 "Small bodies in the solar system" (1987–1992) led to substantial and sustained reinforcement of this field of research in Germany. In addition, decisive impulses for the development of the field of stellar evolution in Germany were provided in recent years by the DFG Priority Programme 471 "The physics of stellar genesis" (1995–2001), which involved numerous institutes. It has been clearly demonstrated that advances in understanding stellar genesis can only be achieved by close cooperation between the theoreticians and the observers.

Thanks to Collaborative Research Centre 301 "The physics and chemistry of interstellar molecular clouds" (1985–1999), the University of Cologne has been in a position, during the last fifteen years, to assume a role in millimetre and submillimetre astronomy which is recognised worldwide. The new Collaborative Research Centre 494 "The evolution of interstellar matter: Terahertz spectroscopy in space and in the laboratory" was established in 2000 at the Universities of Cologne and Bonn, and at the MPI for radio astronomy. It too will provide contributions to the field of stellar genesis. Collaborative Research Centre 328 "The evolution of galaxies" (1987–1998), cemented Heidelberg's position as a centre for extragalactic research, as demonstrated by the establishment there of the new Collaborative Research Centre 439 "Galaxies in the young universe", in 1999. A considerable innovative surge was achieved by the new Munich Collaborative Research Centre 375 "Astroparticle physics" (since 1995). Its interdisciplinary approach also proved a great attraction to students. During the past nine years the Bonn-Bochum Research Training Group 118 "The Magellanic Clouds and other dwarf galaxies" (1992–2001), has fused these two locations, successfully trained a large number of doctoral students and contributed considerably to scientific advancement. These advances form the foundation for the Research Training Group established in 2002 "Galactic clusters as laboratories for baryonic and dark matter". Both the University of Bochum and the University of Bonn are also involved in this college. In addition, the DFG Research Unit "Laboratory astrophysics", recently formed at the Universities of Chemnitz and Jena, must be mentioned in this context.

In the course of the DLR "Space science" programme, the German share of the ESA science programme (currently around €93 million excluding the ISS) and the project sponsorship is funded by the national extraterrestrial research framework (currently around €37 million per annum). Around half of these funds are expended on astronomy. The national programme allows for funding of instrument contributions to ESA missions (for example XMM-Newton, Herschel), international cooperation (for example ROSAT, SOFIA) and last but not least the development of small national missions (for example ABRIXAS), and therefore represents the heart of competitiveness and the capacity for innovation at German research institutes. It therefore provides greater room for action in advancing national interests than the compulsory ESA programme for example. In order to utilise ESA investments optimally (including for the space station, among other things) and especially in order to retain the globally competitive position that the development of technology in German institutes and enterprises enjoys, it is necessary to strike a healthy balance between the national programme and contributions to ESA. Unfortunately, the national extraterrestrial physics programme has shrunk steadily in recent years (by more than 40% since 1996) and will continue to do so in the next few years, so that this balance no longer exists. It is absolutely vital to reverse this trend in the near future in order to retain German space research's capacity for innovation and competitiveness.

DLR national extraterrestrial programme

German astrophysicists were and are active in numerous EU networks and lead a number of them within the framework of the TMR programme, which has appreciably stimulated international collaboration and postdoctoral student exchanges. These EU networks are characterised in particular by a

EU networks

very adequate endowment with travelling and visitors funds. Moreover, the EU's Marie-Curie Programme funds young German scientists for visits of up to two years within Europe, as well as funding the research visits of young foreign scientists to German institutes. The expected expansion of these and similar networks and fellowship programmes in the future is expressly welcomed.

4.5 Training

University astronomy training is almost exclusively restricted to undergraduate physicists. This ensures that astronomers enjoy a solid education in physics and mathematics. Although it is still possible to complete an astronomy doctorate as a first university qualification at individual universities, the opportunity remains practically unused and is not recommended. Astronomy/astrophysics is offered as a subsidiary or optional subject within the physics course at universities hosting astronomical institutes. At some universities astronomy may be taken as a subsidiary subject for the German pre-diploma (intermediate examination). By writing their thesis or dissertation in astronomy, students enjoy an excellent, broad education thanks to the methodical diversity of astronomical research, and do not suffer in any way from poorer career prospects outside of this science compared to other physics students. In addition, astronomers educated in Germany are successful when applying for scientific posts abroad.

The astronomy education offered in lectures, seminars and placements is very well accepted by physics students, and to a lesser degree by chemistry, mathematics and informatics students. Astronomy is a physics "crowd puller". Around 9% of all physics graduate theses are written on astronomical topics (see Table 6.4). This figure should be viewed against the backdrop of only around one third of all universities providing a graduate physics programme actually hosting an institute with a focus on astronomy. Even universities with a large number of physics students often do not offer an astronomy course. This is an alarming fact, especially considering the training of prospective physics teachers – after all, astronomy occupies a central position in our world view in terms of the natural sciences.

The geographical distribution of universities offering astronomy courses displays some peculiarities. Federal states offering astronomy at several locations contrast with those with no astronomy offers whatsoever. These include the states of Saarland and the Rhineland-Palatinate as well as Mecklenburg-Western Pomerania and Saxony-Anhalt. After the planned conversion of the full astronomy professorship in Frankfurt, the state of Hesse will probably also become a blank on the astronomy map (see Figure 6.1).

Young scientists

The dismantling of the scientific middle tier at universities and restrictive German employment legislation hamper the further qualification of young scientists for management positions in science. The restrictions on the duration for which personnel can be employed at fixed term posts can hinder the progress of long-term projects in many cases, worsened by the necessary personnel replacement, and obstruct healthy career planning for many young scientists. The practical effects of the German Framework Act for Higher Education, which came into effect at the beginning of 2002, cannot yet be judged in detail.

Compared to other countries, for example the USA, it is only after much delay that young scientists are in a position to initiate and conduct their own research projects. This failing affects all institutions and necessarily means that it is relatively late before young German scientists come to management positions and can make a name for themselves abroad. The young scientist groups at the Max Planck Institutes and the DFG's Emmy Noether Programme represent useful instruments for counteracting these developments. It will only be possible to assess the consequences of introducing junior professorships at universities in the coming years.

4.6 Astronomy and the public

Public education

Astronomy is highly popular. It is a subject that counts among the crowd pullers when it comes to inspiring young people to study physics. The fascination of research in the natural sciences can be particularly well demonstrated by astronomy, among other things because the questions posed by the origin and evolution of the universe form a central aspect of our world view. Astronomy contributes greatly to public education, in part professionally (via planetariums, museums and exhibitions), in part semi-professionally (for example the popular television programme Alpha-Centauri BR-alpha), in part directly via research establishments (public lectures, open days), public observatories and astronomical societies (there are around 200 societies in Germany with more than 10,000 registered members) and – increasingly of late –via information events for pupils both in their schools and when visiting universities.

A very positive example of this are the school astronomy lessons in a number of the new federal states – an example from which much could be learnt. Concrete efforts are also being made in the old federal states. The University of Göttingen observatory MONET project (Monitoring Network of Telescopes) should be emphasised here. It is funded by the Alfried-Krupp-von-Bohlen-und-Halbach Foundation, which is dedicated to integrating general education schools in observation programmes employing robotic telescopes. Around 40% of observation time will be made available to the schools.

This already large effort should be increased further in future and citizens be brought into even closer contact with astronomy. The most interesting new development in this context is the "virtual observatory". In the presence of suitable infrastructure (for example the public observatories and schools) this would allow many interested young and older people easy access to the fascinating world of astronomy.

Knowledge transfer, cooperation with industry

Astronomy is a high-tech subject. The development of detectors for the various wavelengths places the highest demands on sensitivity and precision. Developments often take place in the respective institute's laboratories and are then awarded to industry for production and further development. Similar conditions apply to telescope development, controls, data recording systems, analysis software, etc. Knowledge transfer is thereby facilitated at all levels:

- training and personnel transfer,
- development and contracts,
- licensing,
- cooperation contracts with industry.

Image processing is one of the most important tools of modern astronomy. However, thanks to developments in this field, knowledge transfer into very different fields to those mentioned above also occurs, for example in medicine – in the early diagnosis of skin cancer, tumour diagnostics, osteoporosis, determination of the depth of anaesthesia, pre- and perinatal diagnostics, ECG analyses, etc. Thanks to its speciality, quantitative image analysis, astronomy has expanded into an important technology driver, even outside of its true field of activity. Further knowledge transfer comes from astroparticle physics – for example photon detectors and cryogenic detectors.

This also leads to a completely different type of knowledge transfer, such as demonstrated by the H.E.S.S gamma telescope system project: the University of Namibia is actively integrated in this project and is thereby given the opportunity to conduct cutting-edge research. The complex steel structure is built by a Namibian company; in this way they also acquire engineering know-how.

The space research field has given rise to numerous technological developments that have directly led to industrial products. Production of the X-ray mirrors for ROSAT, XMM and, in the future, XEUS can be emphasised here. Industrial methods for producing aspherical optics, in particular progressive lenses, have been given significant impulses. The development of innovative X-ray semi-conductor detectors and strip detectors in the low-energy gamma range also offers a great range of potential applications, for example in microstructure analysis, X-ray microscopy and materials diagnostics. In the infrared and optical range "adaptive optics" for ophthalmological diagnostics and surgery applications, infrared spectroscopy for prenatal diagnostics, the development of lightweight silicon carbide mirrors (with special thermal properties for solar observations) and high-precision interferometry are particularly worthy of mention. From the field of radio astronomy come millimetre-sized (MMIC) amplifiers, which work up to 230 GHz and which may play a major role in the future of cellular telephone applications. Other applications have resulted from laboratory astrophysics; for example the development of long-range microscopes and methods for analysing light diffraction in dusty media, which may prove useful in environmental engineering and pharmacy. The C60 molecule ("Fullerene"), so technologically important today, was discovered and produced in useful quantities in astrophysical laboratories (the 1996 Nobel Prize for Chemistry went to Curl, Kroto and Smalley in 1996).

These are only some – and described only briefly – of the most recent knowledge transfer activities in the astrophysics field. This positive energy should be carried into the future. The necessary resources can be acquired from the existing BMBF technology transfer programmes.

5 Recommendations

5.1 Some fundamental aspects[1]

In this chapter the Council of German Observatories (RDS) presents a series of recommendations which, according to their analysis, should be implemented over the next 15 years in order to ensure that German astronomers and astrophysicists remain active at the forefront of worldwide research. The recommendations concern, on the one hand, the funding of projects, experiments and technologies whose scientific necessity was described in Chapters 2 and 3 and, on the other hand, personnel and infrastructure, principally based on the analysis made in Chapter 4. This strategy combines strong position-developing, building and using new instruments, and improved structures and networking. In order to work productively and creatively and to achieve first-class results with the recommended projects/instruments, researchers require sufficient freedoms. In particular, special attention must be paid to the training of junior scientists.

Moderate growth

The recommendations made in this memorandum imply an overall financial volume equivalent to *a moderate total real growth of around 10% (slightly more than 1% p.a.)* of existing funding for the period 2003–2009, based on the year 2000. In the opinion of the Council, growth of this magnitude is both necessary and justifiable based on current and expected developments in the field, expansion in new directions such as astroparticle physics or gravitational wave research and the *absolutely necessary improvements to university infrastruc-*

---

[1] The recommendations given in this memorandum are based on the priorities specified by the Council of German Observatories (RDS) issued at the end of 2001.

ture. However, at the same time, the capabilities and conditions of the sponsors must be taken into consideration in a realistic manner.

The cornerstone of this moderate growth includes a mutual initiative, inspired for this purpose, of the federal states, the DFG and the MPG. It will improve the organisation and infrastructure of the university groups, judiciously develop BMBF cooperative research groups, including astroparticle physics groups, and stabilise and moderate growth in the DLR's national extraterrestrial research budget.

Of fundamental importance to the performance and innovative capacity of German researchers are international participation (also politically very important) and functioning national (competition-oriented) funding instruments. It is recommended that these proven structures be retained and developed further where necessary.
Specifically, these include:

Astronomy funding structures

- ESO (ground),
- ESA (space).

Involvement in international institutions

- BMBF: cooperative ground-based astronomy/astrophysics research,
- BMBF: cooperative astroparticle physics research,
- BMBF: cooperative space research,
- BMBF: linking institutes to high-performance computing centres,
- DLR budget for extraterrestrial research,
- DFG: Priority Programmes, Collaborative Research Centres, Research Units, Research Training Groups, Transregional Collaborative Research Centres, Individual Grants Programme,
- MPG: operation of nationally important observatories and scientific institutes, cooperation with universities and state institutes, establishment of international Max Planck Research Schools,
- Federal States: funding for universities and state institutes,
- Foundations: special funding for projects and scholarships.

National funding

In addition, there are various European Union programmes, in particular the Networks and Fellowships.

These are the reasons why German involvement in the ESO and the ESA science programme are top priority for BMBF and BMBF/DLR funding; they form the basis for a number of important projects. Table 5.1 lists the estimated costs for the BMBF and BMBF/DLR. A German contribution to ESO of €22.5 million p.a. was assumed.

Table 5.1: International involvement.
The figures in this table are based on 2001 and are given in €million. With regard to the ESA contribution it is assumed that 50% of the DLR contribution to the ESA science programme is utilised for astronomy/astrophysics. Only the project-specific costs are listed for all sponsors, but not the institutional funding. Here, and in the following tables, the * character designates that the project/initiative is assigned especially high priority.

		2003–2009 (€million)						2010–2016 (€million)				
G/S	Project	P	States	BMBF	DLR	DFG	MPG z.	P	States	BMBF	DLR	DFG	MPG z.
S	ESA science programme	*			346.0			*			350.0		
G	ESO (VLT, La Silla ....)	*		180.0				*		180.0			

5.2 Instruments and projects

The information discussed in Chapter 2 has already indicated the methods and instruments required to address the central scientific questions of the coming decade. The recommendations for instruments and technologies span the entire range of the electromagnetic spectrum and also include gravitational wave, neutrino and astroparticle physics methods. They are inseparably connected with recommendations for reinforcing theory and numerical simulations. The aim is to nominate the best respective techniques and projects for solving the scientific problems mentioned in the four key topics, in the sense of modern, wavelength- and method-spanning astrophysics. The recommendations aim to achieve a measured balance.

Recommendation priorities

Membership in the international organisations ESO and ESA guarantees astronomers in Germany access to leading global observatories and is therefore continued at the highest priority. Projects run by these international organisations are additionally prioritised here if national financial efforts over and above the normal contribution appear necessary. The following fundamental criteria were assumed when prioritising projects and initiatives:

5.2 Instruments and projects

- considerable progress/improvement of measurement options (Chapter 3) and contributions to the memorandum's core topics (Chapter 2, Table 5.2),
- competitiveness of German researchers in this field,
- balance between the various magnitudes of projects discussed in Chapter 3 (large international projects, national or bilateral initiatives, innovative technologies, etc.),
- coherent interaction between various methods and wavelength ranges,
- consideration of the results of independent expert reporting procedures (for example within the framework of ESA, ESO and DLR projects).

After summarising those projects considered "realistic", or which will probably soon be realistic for the period covered by the memorandum, the critical large projects ("golden bullets") were selected in accordance with the above criteria and a series of important smaller projects and initiatives identified. *Only those "golden bullet" projects that promise an improvement of an order of magnitude or more in terms of sensitivity and/or resolution and offer a fundamental contribution to at least one of the core topics were accepted.* In addition, it was also taken into consideration that some future observatories will be "unique" and without global competition ("global observatories", for example ALMA, JWST), making participation absolutely necessary if Germany does not want to take its leave of an entire branch of astronomical research for many years to come.

The recommendations were then divided into *two priorities*. It is highly desirable for the future development of German astronomy and astrophysics that both the first and second priority projects are realised. However, first priority initiatives (denoted by a "*" in the following tables) are regarded as absolutely necessary in order to maintain the competitiveness of German astronomy and astrophysics. In the Council's opinion, their loss would lead to grave harm to the development of German astrophysical research. Wishes over and above the first two priorities are also mentioned below, but are regarded as lower priority. The recommendations for the instrumentation and technology programme differentiate between *three different financial magnitudes* and between *observatories on the ground and those in space*. The initiatives are also divided into two time periods (0 to 8 years and 9 to 15 years), where detailed forecasts for the latter period are inevitably less precise.

5 Recommendations

However, in addition to the forward-looking projects already identified, room for as yet unknown innovations must be created. This aspect will be addressed by an *"innovation fund"*.

Table 5.2 summarises the instruments and projects for which scientific objectives are required as discussed in the four scientific topics in Chapter 2. The subsequently discussed infrastructure measures are equally crucial to all topics.

Table 5.2: Overview of projects in terms of scientific topics

The initiatives are divided into various categories (international involvement, large, medium and small projects, existing establishments), ground-based projects (G) and space-based projects (S). The x designates the relevance of a project/initiative to a given topic (Sections 2.1 to 2.4), xx indicates that a particularly important or fundamental contribution is expected.

Category	G/S	Project	Topic 1	Topic 2	Topic 3	Topic 4
Intl. involvement	S	ESA science programme	XX	XX	XX	XX
Intl. involvement	G	ESO (VLT, La Silla ....)	XX	XX	XX	XX
Large: >25 €m	S	SOFIA		XX	X	XX
Large: >25 €m	S	Herschel	X	XX	X	XX
Large: >25 €m	S	XEUS	XX	XX	X	X
Large: >25 €m	G	50m+ (e.g. OWL)	XX	XX	X	XX
Large: >25 €m	G	ALMA	XX	XX	X	XX
Large: >25 €m	G	ALMA:APEX		X	X	X
Large: >25 €m	S	Kleinm./bü. Bet.	X	X	X	X
Large: >25 €m	S	DARWIN		X		XX
Med.: 10-25 €m	S	LISA	X	XX		
Med.: 10-25 €m	S	Solar Orbiter			XX	
Med.: 10-25 €m	S	Beneficial space telescope	X	X	X	X
Med.: 10-25 €m	G	Innovation fund	X	X	X	X
Med.: 10-25 €m	G	VLT(I)/LBT	XX	XX	XX	XX
Med.: 10-25 €m	G	H.E.S.S./MAGIC		XX	X	
Med.: 10-25 €m	G	GNO/BOREXINO/LENS	XX	X		
Med.: 10-25 €m	G	SKA	XX	XX	X	
Med.: 10-25 €m	G	ICECUBE	X	X		
Med.: 10-25 €m	S	Solar Orbiter			XX	
Med.: 10-25 €m	G	Astroparticle development	X	X		
Small: <10 €m	S	Planck	XX	X		
Small: <10 €m	G	Access to computing centres	X	X	X	X
Small: <10 €m	G	Virtual Observatory	XX	XX	XX	XX
Small: <10 €m	G	Adaptive optics/detectors	X	X	X	X
Small: <10 €m	G	GREGOR			X	
Small: <10 €m	G	LIGO II/gravitational waves?	X	X		
Small: <10 €m	S	GAIA	X	X	XX	X
Small: <10 €m	S	NGST	XX	XX	X	XX
Small: <10 €m	S	UV (WSO)	X		X	
Small: <10 €m	S	GLAST		XX	X	

Table 5.2: continued

Category	G/S	Project	Topic 1	Topic 2	Topic 3	Topic 4
Small: <10 €m	G	Robotic telescopes	X	X	X	X
Small: <10 €m	G	D.M.:CRESST/GENIUS	XX	X		
Small: <10 €m	G	HET/SALT		X	X	
Existing establ.	S	INTEGRAL		X	X	
Existing establ.	S	SOHO			XX	
Existing establ.	S	XMM-Newton	XX	XX	X	X
Existing establ.	G	Calar Alto	X	X	X	X
Existing establ.	G	Effelsberg/VLBI		XX	X	X
Existing establ.	G	GEO600	X	X		
Existing establ.	G	IRAM		XX	X	XX

A number of instruments are required for work on well-defined special topics. CRESST and GENIUS, for example, serve in the search for dark matter particles, Planck for surveying the microwave background and GREGOR for studying the sun. On the other hand, some "workhorse" observatories cover the entire range of astrophysical research. These include the large, ground-based telescopes, for example, but also several space-based telescopes.

Specialised instruments and "workhorses"

5.3 Large projects

In the field of large space-based projects, Herschel, SOFIA and the technological preparation and instrumentation for XEUS are first priority tasks (Table 5.3) for the period 2003–2009. At least one small mission or bilateral project, such as MEGA, PRIME, ROSITA or SUNRISE, for example, is highly desirable, similar to the technological preparation for DARWIN. In addition, instrumentation for the DARWIN mission is a first priority for the period 2010–2015.

In the field of large ground-based projects, the ALMA project (and its precursor programme APEX) is assigned first priority for the period 2003–2009. The next generation of ground-based optical/IR telescopes (50 m or more in diameter, such as the OWL project, for example) is at the forefront for the period 2010–2015. For ALMA, and generally also for 50m+/OWL, it is assumed that building can be financed by a constant (in real terms), or slightly increased, German contribution to the ESO budget, although "in-kind" contributions via IRAM and the MPG are also important.

Table 5.3: Large projects (> 25 €m for at least one of the periods)

The figures in this table are in millions of euros at 2001 prices and must therefore be inflation-corrected for the appropriate year. The basic funding for the institutes and universities is not separately listed in the table. For the MPG only the additional budget funds directly required for the projects (ad.) is listed. First priority projects are denoted by the * character.

| G/S | Project | Type | P | 2003–2009 (€million) ||||| Remarks |
				States	BMBF	DLR	DFG	MPG ad.	
S	SOFIA	Operation+instrum.	*			25.0	1.5	1.0	DLR SOFIA inst., DFG/MPG instrm.
S	Herschel	Instr. PACS, HIFI	*			30.0			
S	XEUS	Instr.	*			10.0			
S	XEUS	Technology	*			4.5			Support from DLR, in part from ESA
S	Kleinm./bil.Bet.	Satellite/instr.	*			25.0			e.g. MEGA/PRIME/ROSITA/Sunrise
S	DARWIN	Techn./instr.	*			2.0			
G	ALMA:APEX	Building APEX	*						
G	ALMA	Building APEX	*		[20]			3.0	Included in BMBF:ESO+MPG:IRAM
G	ALMA:APEX	Use of APEX			1.5				
	Priority 1 total					69.5	1.5	4.0	75.0
	Priority 2 total				1.5	27.0			28.5

| G/S | Project | Type | P | 2010–2016 (€million) ||||| Remarks |
				States	BMBF	DLR	DFG	MPG ad.	
S	SOFIA	Operation+instrum.	*			25.0			DLR SOFIA institute
S	Herschel	Instr. PACS, HIFI	*			4.5			
S	XEUS	Instr.	*			30.0			
S	Kleinm./bil.Bet.	Satellite/instr.	*			25.0			e.g. MEGA/PRIME/ROSITA/Sunrise
S	DARWIN	Techn./instr.	*			30.0			
G	50m+ (e.g. OWL)	Build	*		10.0				
G	ALMA	Build	*		[5]	89.5			Included in BMBF: ESO+MPG:IRAM
	Priority 1 total								99.5
	Priority 2 total				10.0	25.0			27.5

5.4 Medium-sized projects

In the field of medium-sized space-based projects, DLR-funded use of space telescopes ("DLR cooperative research") and the instrumentation for the Solar Orbiter and LISA are first priority for the period 2003–2009 (Table 5.4). An increase in the Max Planck contribution to the Solar Orbiter would be highly desirable for this period. Because instrumentation-building for the Solar Orbiter and LISA will probably extend into the 2010–2015 period, these two projects remain first priority, similar to DLR cooperative research. If LISA can be realised earlier by means of cooperation between ESA and NASA, it would be desirable to bring forward the German contribution correspondingly.

In the field of medium-sized ground-based projects, the astroparticle projects H.E.S.S./MAGIC and ICECUBE, GNO/BOREXINO/LENS, the VLT(I)/LBT instrumentation and Max Planck participation in the technological preparation for the Square Kilometre Array are viewed as first priority for the period 2003–2009. This includes the proposed FrInGe interferometry data centre. Another project is the innovation fund mentioned above, supported by BMBF cooperative research, the MPG and the federal states. The fund makes it possible to include new, exceptionally high-quality developments, as yet unforeseeable by this memorandum. Further development of cooperative research into astroparticle physics is also desirable for this period, together with more funds for VLT(I) and LBT. The priorities for the period 2010–2015 are very similar to those for the first period, although the SKA building and instrumentation phase will be supported by BMBF cooperative research and the MPG. ICECUBE will by then have reached the use phase.

5.5 Small projects

In terms of minor involvement in space projects, the Planck data centre and the GAIA data centre have first priority for the period 2003–2009 (Table 5.5). It is hoped that the federal states will participate in the GAIA data centre. Hardware participation and collaboration in GLAST, a German hardware contribution to the JWST (MIRI), involvement in GAIA instrumentation and in developing the technology and instrumentation for a future

Table 5.4: Medium-sized projects (10 to 25 €m for at least one of the periods)

The figures in this table are in millions of euros at 2001 prices and must therefore be inflation-corrected for the appropriate year. The basic funding for the institutes and universities is not separately listed in the table. For the MPG only the additional budget funds directly required for the projects (ad.) is listed. First priority projects are denoted by the * character.

2003–2009 (€million)

G/S	Project	Type	P	States	BMBF	DLR	DFG	MPG ad.	Remarks
S	LISA	Instr.	*			10.0			
S	Solar Orbiter	Instr.	*			10.0			
S	Space telescopes	Use	*			13.0			
S	Solar Orbiter	Instr./use				3.0		1.0	
G	VLT(I)/LBT	Instr.	*		11.0			2.5	incl. FrInGe
G	Innovation fund	Techn./instr.	*	5.0	5.0			5.0	New, excellent small initiatives
G	H.E.S.S./MAGIC	Build/use	*		9.2			2.5	
G	Astroparticles	Build/use	*		8.8			2.0	
G	ICECUBE	Build/use	*		8.8				
G	VLT(I)/LBT	Instr.						5.0	
	Priority 1 total			5.0	42.8	33.0		12.0	92.8
	Priority 2 total					3.0		6.0	9.0

2010–2016 (€million)

G/S	Project	Type	P	States	BMBF	DLR	DFG	MPG ad.	Remarks
S	LISA	Instr.	*			10.0			
S	Solar Orbiter	Instr.	*			10.0			
S	Space telescopes	Use	*			15.0			
S	Solar Orbiter	Instr./use				2.0			
G	VLT(I)/LBT	Instr.	*		3.0			2.0	incl. FrInGe
G	Innovation fund	Techn./instr.	*	5.0	5.0			5.0	New, excellent small initiatives
G	H.E.S.S./MAGIC	Build/use	*		10.0			1.5	
G	Astroparticles	Build/use	*		2.5			2.0	
G	SKA	Techn./instr.	*		10.0			5.0	
G	Astroparticle development	Instr.	*		13.0				
G	VLT(I)/LBT	Instr.			5.0			2.0	
	Priority 1 total			5.0	34.0	35.0		10.5	84.5
	Priority 2 total				15.0	2.0		7.0	24

5.5 Small projects

UV space-based telescope (such as the WSO for example) are all desirable. Involvement in the MIRI instrument for the JWST represents a special case here, because it may then be possible to realise a significant German contribution to the JWST, which is very important to the German research community. Overall, the JWST is therefore definitely categorised as a first priority project and it can only be hoped that a German hardware contribution can be realised even if the financial situation is difficult. Differentiation between first and second priority was purposely avoided for the 2010–2016 period, because it is not possible to estimate requirements exactly at this time. The most important projects during this period include Planck data evaluation and contributions to GAIA, and possibly UV/WSO instrumentation.

In the field of small, ground-based projects the GAVO project, access to high-performance computing centres, participation in gravitational wave experiments (LIGO II), technology development in the field of adaptive optics and new detectors, and building the solar telescope GREGOR all have first priority for the period 2003–2009. In addition, BMBF funding of dark matter experiments (CRESST, GENIUS), an extension of the adaptive optics/detectors technology programme and participation in LIGO II, as well as funding of new, robotic telescopes and HET/SALT, are all desirable. The long-term elements continue to be included in the 2010–2016 period. Here, too, first priorities remained undefined intentionally.

Table 5.5: Small projects (< 10 €m in both periods)

The figures in this table are in millions of euros at 2001 prices and must therefore be inflation-corrected for the appropriate year. The basic funding for the institutes and universities is not separately listed in the table. For the MPG only the additional budget funds directly required for the projects (ad.) are listed. First priority projects are denoted by the * character.

G/S	Project	Type	P	States	2003–2009 (€million) BMBF	DLR	DFG	MPG ad.	Remarks
S	Planck	Use	*			2.0			
S	GAIA	Instr./use	*	5.0		5.0			
S	NGST	Instr.: MIRI				10.0			
S	UV (WSO)	Techn. + instr.				5.0			
S	GLAST	Instr. + use				2.5			
G	Access to computing centres	Instr.	*	5.0					
G	Virtual Observatory	Instr.	*		2.5				
G	Adaptive optics/ detectors	Technology	*						
G	GREGOR	Build	*	4.0	3.5				University build funding
G	LIGO II/ gravitational waves	Instr. + techn.			5.0				
G	Robotic telescopes	Instr. + use		2.0	2.0				Additional funds from foundations
G	D.M.:CRESST/GENIUS	Techn. + use			2.5				
G	Adaptive optics/ detectors	Technology			3.5				
G	HET/SALT	Instr. + use			1.0			2.0	Additional funds from VW foundation
	Priority 1 total			14.0	6.0	2.0		2.0	
	Priority 2 total			2.0	14.0	22.5			

(continued on page 191)

Table 5.5: continued

G/S	Project	Type	P States	2010–2016 (€million) BMBF	DLR	DFG	MPG ad.	Remarks
S	Planck	Use			1.0			
S	GAIA	Instr./use	5.0		5.0			
S	UV (WSO)	Techn. + instr.			5.0			
G	Access to computing centres	Instr.	5.0					
G	Virtual Observatory	Instr.		2.5			2.0	
G	Adaptive optics/ detectors	Technology		4.0				
G	LIGO II/ gravitational waves	Instr. + techn.		10.0				
G	Robotic telescopes	Instr. + use	2.0	2.0				Additional funds from foundations
G	D.M.:CRESST/GENIUS	Techn. + use		4.0				
G	HET/SALT	Instr. + use		0.5				
	Total		12.0	23.0	11.0		2.0	48.0

5.6 Organisational instruments

In the first two chapters of this memorandum we described how astronomy is currently in a phase of rapid development. On the one hand, this is the result of considerable technological advances; on the other, the view that astrophysics represents a substantial component of physics research and that the universe can be regarded as a unique and valuable physics research laboratory has been strengthened by the scientific discoveries of the last decade. Examples include the discovery of dark matter and dark energy, and the solar neutrino problem, which has gone far to promote particle models beyond the standard model.

In contrast to other countries this "golden phase" of astrophysics was not taken into account by a commensurate increase in the personnel employed in astronomy (see Chapter 4); instead, established posts were substantially decreased in western Germany (see Table 6.3). Because membership subscriptions to the international organisations ESO and ESA are calculated on the basis of the gross national product of the participating countries, but the number of astronomers in Germany is much smaller in terms of gross national product compared to other countries, the scientific yield of the investments made can almost certainly be noticeably increased by improving the personnel situation.

Strengthening of basic university funding

While the Max Planck Institutes (MPIs), institutes of the federal states and universities were affected by personnel reductions, the latter noticed the effects much more dramatically. The institutional recommendations therefore concentrate on university institutes and institutes owned by the federal states. Many of the problems faced by the university institutes are not astronomy-specific. However, this branch of science is characterised by only around 30% of established scientific posts being available at universities (see Table 6.2). The universities therefore find themselves in an unfavourable position in the competition for the best researchers.

In order to counteract the structural weakness of the university astronomy institutes, *basic institutional funding must be reinforced*, in terms of both personnel and instrumentation aspects. The following remarks also apply, albeit to a lesser degree, to the state-owned institutes.

Increasing personnel

The reductions in established posts at many universities have left considerable holes in the institutes' personnel. After an

5.6 Organisational instruments

initial phase of staff consolidation, the pain threshold has been exceeded at a number of institutes over the past few years: an increasing number of tasks related to training, university self-administration, sourcing of external funding and public relations work must be managed by increasingly fewer personnel. The resulting continuous decrease in research freedom at university institutes is demotivating and therefore also damaging to teaching; moreover, the competitiveness of university research increasingly suffers. It is therefore urgently recommended that this long-term trend be reversed and the discrepancy between established posts and infrastructure on the one hand, and external funding on the other, be at least moderated. In terms of personnel and funding, attempts must be made to bring the support for astrophysics at universities to a level corresponding to that in similar countries (see Chapter 4). Because astrophysics forms an integral component of physics, and is currently one of the most rapidly developing fields within physics, with excellent prospects for fundamental discoveries, it is recommended that this be taken into account in the planning perspectives of physics faculties, in order to counteract the significant underrepresentation (in international terms) of astrophysics.

Strengthening instrumentation

Even if it is assumed that the peak demand for computing capacity can be satisfied by both the existing and the planned high-performance computing centres, suitable hardware is still required by the individual institutes. Of course, this applies equally to working groups that are not active in numerical astrophysics. For example, during data reduction the amount of data accrued is sometimes so great that the smaller institutes are almost doomed to failure simply due to data storage and backup problems. This problem is continuously worsening. Take optical cameras as an example: within just a few years (for example at the ESO) they have grown from $1,000^2$ (SUSY at the NTT) to $16,000^2$ (OmegaCAM on the VST) image elements. In addition, in many places the personnel reductions have led to the care of computer systems having to be carried out by students; in other words, there are no system administrators. It is therefore necessary that the university institutes are better equipped with computers, the appropriate peripherals (storage, backup systems) and system administrators if required. According to the DFG Guidelines these form part of the infrastructure of the institutes and are therefore the responsibility of the respective federal states. Moreover, the workshops at many university institutes have been severely weakened by the staff cuts. Because these workshops repre-

5 Recommendations

sent a necessary requirement for participation in instrumentation projects, this reduction in personnel is directly coupled to a weakening of competitiveness in terms of instrument building. In order to allow university institutes to continue their activities in building and instrumentation, a return to efficient workshops is recommended, which is also desirable from a student education perspective.

Access to optical telescopes

Thanks to the MPG participation in Calar Alto, German astronomers also have access to optical telescopes, beside those available through ESO. Approximately half of the observation time available to Germany is utilised by MPIs, the other half is available to RDS member institutes. The reduction in German participation in Calar Alto, considered necessary by the MPG, will involve a painful restriction in access to optical telescopes for university researchers, especially in the northern hemisphere. While the institutes involved in the LBT operating company will have alternative observation options thanks to the LBT, they will not be available to the universities. Telescope access beyond that provided by the ESO options is necessary for executing large and long-term programmes. This access is guaranteed in countries similar to Germany (France: Franco-Canadian-Hawaiian Telescope; The Netherlands: La Palma; Italy: Telescopio Nazionale Galileo, UK: La Palma, Gemini, Anglo-Australian Telescope). It is therefore recommended that options for acquiring and securing university participation in optical observatories (such as La Palma, for example) be investigated.

Access to the most important electronic publication media

The increased prices of journals, in conjunction with what are at best only stagnating institute budgets, has torn gaping holes in university institute libraries in recent years. The subscription rate for the four most important astronomical journals alone amounts to around 9,700 euros, representing a considerable fraction of any institute's budget. Interdisciplinary and physics journals are therefore only rarely found in astronomy libraries. With regard to the rapidly expanding electronic distribution of data and journals, it is important that university groups have unrestricted general access to the most important electronic publication media and electronically accessible journals. For example, more use could be made of the central journal subscription option as it is used within the MPG. It is recommended that a similar model be aimed for at least at the federal state level.

5.7 Interaction and cooperation

Because of advances in instrumentation and the ensuing moves towards large instruments, astronomy is increasingly becoming a science of large observatories and research cooperatives, which characterise the international competition situation to an increasing degree. Examples in the USA include the Center for Astrophysics in Cambridge, Massachusetts and the seven universities of the University of California and Caltech amalgamated in the Keck Foundation/CARA. Other examples include CNRS/INSU in France, and Cambridge University and PPARC in the United Kingdom. These establishments coordinate their larger research projects on different levels – from observatory instrumentation and the planning and implementation of observation programmes, to theory and numerical simulations. It is the coupling of quality and size (typically ten to twenty chairs) that facilitate synergy, efficiency and success. This is naturally associated with access to the most modern telescopes.

In Germany, non-university institutes can only just (partly in cooperative efforts) keep up with the competition; the universities generally cannot. Positive exceptions are also known, thanks to targeted funding measures such as Collaborative Research Centres and cooperative research, showing that the situation need not remain like this. However, it is conceivable that the growing expense of top-class research in the important forward-looking topics will necessarily also lead to larger research cooperatives if Germany wants to preserve a position as one of the world's leaders. *Larger research cooperatives*

The close cooperation between institutes represents an active research element and one that will be reinforced in the future. It will be further reinforced by the "thematic networks" concept recommended below. However, this should not cause the importance of the many smaller projects to be overlooked. Even today, very small groups can carry out decisive theoretical work, for example, or small, but highly targeted, observation programmes. Flexibility is often the key to success for these small projects. Here, too, the structural weakness of the universities is noticeable, for example in their lack of travelling and visitors funds.

In order to take into account future developments towards larger research cooperatives it is recommended that astronomy and astrophysics be thematically structured to suit current *"Vertical" science structure* scientific requirements. Beside the "horizontal"-oriented technology structure (observatories, instruments), the "vertical" science structure, which focuses mainly on scientific topics, will also be improved.

The task of these networks is to improve the structure of German astronomy by:

- integrating the small institutes in large, internationally competitive research projects and by providing support in the setting up of new, innovative research establishments, including in the universities,
- bundling the expertise distributed across various establishments for mutual research projects,
- greater assertion in international projects,
- coordinating and planning research projects,
- coordinating teaching and teaching export to locations where astronomy is not represented as a subject,
- the exchange of scientists, doctoral students and graduate students between institutes, as well as by organising workshops and summer schools, and
- a science-oriented representation of astronomy in the public eye.

Because these networks must grow organically, development in stages is proposed. In a first phase a "coordination network" will be launched, in which the concepts for the development of thematic-oriented networks and their organisational structure are compiled, in particular in terms of funding options. The development of the thematic networks can start in a second phase. Each institute can join one (or more) of these thematic networks – depending on scientific interest and expertise. For example, the networks can initially orient themselves around the four priority topics in this memorandum (see Chapter 2). In addition, networks can form to take advantage of particularly innovative technological fields such as numerical astrophysics, processing large data records, adaptive optics or the "virtual observatory". The structure of these networks must be flexible; for example, geographically close institutes could form stronger sub-nodes.

Mutual efforts

Funding these networks requires a mutual effort by the federal states, the DFG, the BMBF and the MPG. Due to the integration of numerous smaller establishments encompassing an enormous amount of expertise, but which would not alone be capable of achieving "critical mass", the cost-benefit ratio of such a network will be greater than would be achievable by funding a similar number of individual projects. Together with other funding measures, for example through the DFG, networks will be given the flexibility to use their funds; however, they differ from the DFG Priority Programmes in their strong emphasis on infrastructure effects. It is planned to coordinate the networks through the Council of German Observatories.

Theoretical astrophysics

The traditionally strong German role in theoretical astrophysics must be maintained. This applies both to the field of numerical astrophysics and to more analytical theoretical research. Care should therefore be taken that the fantastic advances in instrumentation achieved in astronomy and the associated commitment of resources do not lead to restrictions in the funding of theoretical work. Only a healthy balance and cooperation between theory, observation and experiment guarantees new discoveries. A balance between these fields is also indispensable for the education of junior scientists.

Increasing the number of subjects

It was shown in Chapter 4 that astrophysics is only represented at around one third of the universities that educate physicists. This means that a huge chance is missed; that of strengthening physics teaching at other universities as a result of the particular attractiveness of astronomy as seen from the students' viewpoint. *The tendency to train physics teachers with no recourse to astronomy is particularly worrying.* In order to alter this situation noticeably and secure nation-wide astronomy teaching, two different, mutually compatible models are proposed:

- Establishing new astronomy institutes or groups. The establishment of astronomy professorships at universities where physics is taught would not only expand the number of courses offered, but also contribute to the strengthening of university-based astronomy research. To ensure competitiveness and in order to win excellent scientists to fill the positions available, care should be taken that these new groups are not too small. It is noteworthy that in the United Kingdom, for example, several new astronomy-teaching locations have been established in recent years.

- Teaching export. University teachers at the large astronomy-teaching locations provide astronomy teaching at neighbouring universities, as is already practised on a small scale. However, this kind of teaching export cannot be achieved with only the existing university teaching staff; new professorships therefore need to be established at the existing astronomy-teaching locations.

The advantage of teaching export is that the newly appointed professors and their assistants are integrated in already active institutes and further splintering of astronomical research can therefore be avoided. However, teaching export can only be realistically implemented at geographically close universities; in order to provide teaching nationwide it would need to be organised and funded at the federal level. It is therefore recommended that a mix of the two models be implemented.

5.8 Securing and reinforcing funding instruments

Basic institutional funding

A number of elements must be associated with personnel and infrastructure. The first important element is basic institutional funding: from the MPG for its institutes, from the BMBF and federal states for the institutes of the Leibniz Society and from the federal states for the universities and state institutes. The second element includes funding for R&D projects by the DFG and the use of large instruments and space-based telescopes by the cooperative researchers of the BMBF and DLR. Increasing use has been made of funding by the European Union in recent years, for example in the shape of international networks. Finally, general employment and service regulations need to be considered.

Funding young scientists

One of the structural weaknesses, and a competitive disadvantage, of German astronomy is that a researcher only assumes responsibility relatively late in his career, as well as the lack of long-term, personal academic development perspectives for young astrophysicists (similar to physicists in general). While in recent years the number of doctorates in the West has increased by around 50%, there has been a simultaneous decrease in established posts by around 25% (see Table 6.4). This has led to a dramatic deterioration in career opportunities for young scientists. *It is absolutely vital to make research and teaching more attractive to excellent young scientists in the coming years.* In order to achieve this the Council of German Observatories recommends a number of measures:

5.8 Securing and reinforcing funding instruments

- reinforcing basic university funding, as clearly described above,

- intensifying cooperation between Max Planck Institutes, state institutes and universities. The MPG and DFG system evaluations pointed out the competitive gradient between the Max Planck Society and the universities. In order to promote the independence of young scientists and to strengthen cooperation between the universities and the MPG, it is recommended that the MPG establish research and young research groups at universities. It is assumed here that a long-term perspective will be compiled for these groups of researchers in cooperation between the MPG and the respective universities. With regard to astrophysics, these cooperation instruments could accelerate the badly needed broadening of the astrophysical base at the universities. This is required most urgently in the new federal states.

- Tenure Track model for excellent junior scientists. The introduction of a "Tenure track model" would represent a significant improvement. It would provide excellent young scientists with a visible long-term perspective at an early stage. In principle, the junior professorships envisaged by the new (German) Framework Act for Higher Education provide for this option and are therefore welcomed. However, at the current time it still cannot be seen how the individual states will apply this instrument and whether it will provide a noticeable improvement to the situation of the new generation of scientists. Alternatively, the introduction of a Tenure Track model should be developed by the universities in cooperation with the DFG and the MPG (working title: Schwarzschild programme). It would complement the DFG's Emmy Noether Programme: for example, the universities could achieve continuance of one of the MPG junior research groups recommended above, and thus effectively of the external funding of teaching activities for a period of several years, if they commit themselves to providing a long-term position for the group head (if success has been demonstrated). After inviting applications, selection should made by an appointment committee similar to the case for associate professor university positions and heads of MPG junior research groups.

- Establishment of interdisciplinary chairs and centres. Astrophysics radiates strongly into other subjects and is therefore eminently suitable for establishing interdisciplinary chairs

5 Recommendations

and centres. For example, consider chairs for plasma and astroparticle physics, or centres for stellar and planetary genesis, gravitational wave astrophysics, astrochemistry and cosmology.

Project funding

The funding of research projects by the DFG and cooperative research are essential elements of the German research landscape, in addition to European Union funding measures and the support provided by foundations. This external funding is invaluable for maintaining research at universities, but is also vitally important for state institutes and Leibniz Society institutes. The various funding programmes differ in their extent and duration; while DFG Collaborative Research Centres offer long-term funding perspectives on the one hand, other funding instruments are typically restricted to two or three years duration.

(Transregional) Collaborative Research Centres and Priority Pogrammes

With regard to the large projects discussed above and the increasing importance of complex modelling, theoretical interpretation and numerical simulations, synergy effects will increase in relevance during the next decade. One urgent recommendation is therefore that the research community should undertake extra efforts to apply for Collaborative Research Centres and Priority Programmes and that the DFG become increasingly capable of supporting these funding instruments. The establishment of Transregional Collaborative Research Centres is welcomed. Due to their long-term perspective, these funding instruments offer an opportunity for positive research planning, with particular relevance to personnel planning.

"Rolling Grants"

The integration of research groups into large, often international cooperatives and into longer-term projects often does not suit the concept of biannual project funding upon which the DFG Individual Grants Programme, for example, is generally based. Astrophysical research does not usually take place in the framework of isolated or easily separable biannual projects. On the other hand, the active groups are usually successful in acquiring extensions to, or follow-up, project funding. The main problem here is the lack of ability to plan ahead: the approval of an application for extension is only made public shortly before the original programme expires. As a consequence of this, experienced staff often cannot be retained or suitable new staff be sourced at short notice.

It is therefore recommended that so-called "Rolling Grants", such as already exist in other countries (for example in the United Kingdom), be established as funding instru-

5.8 Securing and reinforcing funding instruments

ments. A research group or collaboration project is approved for a period of four years. An evaluation after around two years will then decide whether project continuation past the four year mark is justified, whether funds should be reduced, or even increased, after four years, or whether the project should terminate after the four year period. In case of a two year extension, the next evaluation will be in two more years and so on. Using this model the funded research group always has two years planning security, allowing timely application invitations to be published and the most suitable international researchers to be sought. The Council of German Observatories is convinced that such research funding represents a considerably more efficient use of funds, for a comparable expenditure, than the previously customary practice of biannual projects.

Acceleration of external funding

The problem of the lack of staff flexibility at smaller institutes and their dependence on external funding is often highlighted further by long processing times and decision deadlines. It would therefore be highly desirable, especially with regard to smaller applications, to shorten these deadline dates considerably and thereby to take the life and career plans of young scientists into consideration. This would also enhance the competitiveness of the research institutes when seeking new, or reemploying existing, scientists.

Utilising European programmes

Astronomy and astrophysics are international in character. The field is therefore in an excellent position to continue to utilise European programmes in the course of European integration and to increasingly take up the opportunity for international study courses in the future. It can also promote young scientists as encouraged by the Erasmus programme or the Marie Curie Fellowships. In this context, the most recent DFG (Emmy Noether Programme) or MPG initiatives (International Research Schools) are welcomed and will continue to remain desirable in the future.

Reinforcing research using top-class telescopes

German astronomers are successful in acquiring observation time at the heavily over-booked top-class observatories (for example by a factor of around 7 for the Hubble Space Telescope). Their success in the competition for this telescope time, which is awarded in an international peer-review process, attests to the excellent quality of a given project. In order to present a maximum of scientific results from the acquired data, human resources must be secured. It is therefore recommended that successfully acquired observation time on the top-class telescopes be more strongly coupled to appropriate personnel

5 Recommendations

Creating options for leave of absence

funding even for observatories not currently involved in cooperative research (for example Chandra).

The poor representation of German scientists in the management levels of international organisations such as ESO and ESA has its root cause in the difficulties involved in the long-term release of top German researchers from their other duties. Creating options for leave of absence would be much simpler if long-term teaching stand-ins could be funded by suitable measures. On the one hand, funding teaching stand-ins for extended periods (around 3 years), for example by means of a new and targeted research prize programme, would give senior university teachers sufficient time and freedom to initiate new research establishments and scientific collaborations (for example in the planned networks). On the other hand, it would allow them to be available for tasks in international organisations and thus to increase the influence of German scientists within these organisations. Research prizes can be awarded for projects in cooperation with other institutes and should include travelling funds and non-monetary resources in order to allow regular visits to the partner institutes (for example several days each month).

Promoting women

The proportion of women in astronomy is horrifyingly low. This also applies to physics in general. Internationally, the proportion of women in Germany is just ahead of countries such as Japan, Switzerland and India among IAU members; but at less than 4%, Germany takes its place behind practically all other countries (United Kingdom 7%, Netherlands 8%, USA 9%, Italy 17%, France 26%; IAU average: 10.5%). While a number of causes are responsible for this low proportion of women, service regulations also play a decisive role. Service regulations do not allow bridging assistance, for example for maternity leave and raising children in young families, especially for women in fixed-term posts. Political support for the planned reforms to service regulations (scientists collective agreement) is therefore urgently required.

Graduate funding

The current and foreseeable mid-term dearth of physicists should give rise to increased funding efforts for doctoral students. In particular, it would be desirable to unify the various funding programmes for doctoral students so that the existing large differences in financial support (ranging from approx. €620 state scholarship up to full BAT-IIa posts – BAT is the *Bundesangestelltentarif* – the (German) Federal Employees' Collective Salary Agreement) are ironed out and adjusted to

5.8 Securing and reinforcing funding instruments

a sensible level (at least BAT IIa/2). Nobody can understand why doctoral students at the same institute, working on similar topics, are paid so very differently; this creates unnecessary conflicts between working groups.

Summary

The most important recommendations for the organisational, personnel and infrastructure fields are summarised below (see Table 5.6):

- noticeable improvements in the infrastructure of the university institutes, especially in personnel,
- establishing new chairs (in particular interdisciplinary, for example in plasma- and astroparticle physics, stellar and planetary research, gravitational wave astrophysics, astrochemistry and cosmology),
- expanding the teaching offered in order to achieve nationwide astronomy training, especially for prospective physics teachers,
- increased use of DFG Collaborative Research Centres, Priority Programmes, Research Training Groups and Transregional Cooperatives, as well as research centres, in particular in the observation, theory and numerical modelling fields,
- launching federal thematic "networks" or cooperatives,
- developing competence centres for technically demanding fields,
- establishing mutual university and MPG research and junior research groups, as well as increased utilisation of the DFG's Emmy Noether Programme for promoting the independence of young scientists and reinforcing cooperation between universities and the MPG,
- establishing Tenure Track positions (working title: Schwarzschild scholarships) at universities and
- simplifying longer-term deputations and leave-of-absence for research through a prize sponsor programme.

Table 5.6: Infrastructure measures

The figures in this table are in millions of euros at 2001 prices and must therefore be inflation-corrected for the appropriate year. The basic funding for the institutes and universities is not separately listed in the table. For the MPG only the additional budget funds directly required for the projects (ad.) are listed. First priority projects are denoted by the * character. The €m 25 for the centres (not included in the total) is intended for the DLR-funded SOFIA institute.

Project	2003–2009 (€million)						2010–2016 (€million)					
	P	States	BMBF	DLR	DFG	MPG ad.	P	States	BMBF	DLR	DFG	MPG ad.
Jun./working/research groups	*				10.0	8.0					10.0	5.0
SFB/Prior./Trans/Res Train/Ind	*				76.0		*				80.0	
Networks on topics 1-4	*	12.0	10.0	2.5			*	12.0	5.0	2.5		
Tenure Track positions		8.0						5.0				
New chairs		8.0						13.0				
Priority 1 total		12.0	10.0	2.5	86.0	8.0		12.0	5.0	2.5	80.0	5.0
Priority 2 total		16.0						18.0			10.0	
Priority 1 overall		118.5						99.5				
Priority 2 overall		16.0						33.0				

5.8 Securing and reinforcing funding instruments

Table 5.7 summarises the initiatives proposed here and lists the resources apportionable to the various sponsors according to priorities and time periods. Similar to the other tables, Table 5.7 includes the estimated costs in €m for the reference year 2001. It is assumed that the budget will be corrected for inflation in the future. The existing institutional funding is included for the federal states (universities and state institutes), the BMBF (Leibniz Society institutes) and the MPG. (It amounts to approximately €110 million p.a. including overheads). With regard to the MPG, the "MPG ad." (additional) column indicates the additional resources required in conjunction with the projects recommended here, which fall outside of the framework of existing institutional funding. "DLR" here refers to the national extraterrestrial programme funded by the BMBF and German involvement in the ESA science programme. With regard to the DLR (ESA), half of the science programme is included, which corresponds approximately to the share of the astronomy/astrophysics research in this programme.

Distribution among sponsors

Table 5.7: Distribution among sponsors

The Max Planck Society has approved current (including for the year 2000) additional funding programmes for astronomy (appointment funds, construction activities, etc.). They are not listed here. The federal states are similar.

Sponsor	All in €m				Annual funding 2003–16 Priority 1	Annual funding 2003–16 Priority 1+2	Annual funding As at 2000
	2003–2009 Priority 1	2010–2016 Priority 1	2003–2009 Priority 2	2010–2016 Priority 2			
DFG	86	80		10	12	13	9
BMBF (ESO)	180	180			26	26	22
BMBF (national)	112	106	16	38	16	19	11
States	434	422	18	30	61	65	58
DLR (ESA)	346	350			50	50	50
DLR (national)	107	130	53	38	17	23	16
MPG (additional)	26	11	6	15	3	4	0
MPG (institutional)	414	423	7	9	60	61	60
Total:	1,705	1,702	99	140	244	260	226

6 Annex

Member institutes of the German Council of Observatories

Dr. Remeis-Sternwarte Bamberg
Astronomisches Institut der Universität
Erlangen-Nürnberg
Sternwartstraße 7
96049 Bamberg
Tel. 0951-95222-0
Fax: 0951-95222-22
http://a400.sternwarte.uni-erlangen.de

Zentrum für Astronomie u. Astrophysik
der Technischen Universität Berlin
Hardenbergstraße 36
10623 Berlin
Tel. 030-314-23783
Fax: 030-314-24885
http://www-astro.physik.TU-Berlin.de

Institut für Planetenforschung
Deutsches Zentrum für Luft-
und Raumfahrt
Berlin – Adlershof
Rutherfordstraße 2
12489 Berlin
Tel. 030-67055-300
Fax: 030-67055-303
http://www.dlr.de/pf/

Astronomisches Institut
der Ruhr-Universität Bochum
Universitätsstraße 150 / NA7-67
44780 Bochum
Tel. 0234-322-3454
Fax: 0234-32-14169
http://www.astro.ruhr-uni-bochum.de

Theoretische Weltraum und Astrophysik
Theoretische Physik IV NB 7/56
Ruhr-Universität Bochum
Universitätsstraße 150
44780 Bochum
Tel. 0234-32 220-32
Fax: 0234-32-14177
http://www.tp4.ruhr-uni-bochum.de

Argelander-Institut für Astronomie
der Universität Bonn
Abteilung Sternwarte
Auf dem Hügel 71
53 121 Bonn
Tel. 0228-73-3655
Fax: 0228-73-3672
http://www.astro.uni-bonn.de/~wcbstw

Argelander-Institut für Astronomie
der Universität Bonn
Abteilung Radioastronomie
Auf dem Hügel 71
53121 Bonn
Tel. 0228-73-3658
Fax: 0228-73-1775
http://www.astro.uni-bonn.de/~webrai

Argelander-Institut für Astronomie
der Universität Bonn
Abteilung Astrophysik
Auf dem Hügel 71
53121 Bonn
Tel. 0228-73-3676
Fax: 0228-73-4022
http://www.astro.uni-bonn.de/~webiaef

Jacobs University Bremen gGmbH
Campus Ring 1
28759 Bremen
Tel. +49 (0)421 200-40
Fax: +49 (0)421 200-4113
http://www.jacobs-university.de/

Max-Planck-Institut
für Radioastronomie
Auf dem Hügel 69
53121 Bonn
Tel. 0228-525-0
Fax: 0228-525-229
http://www.mpifr-bonn.mpg.de

Lohrmann-Observatorium und
Professur für Astronomie im
Institut für Planetare Geodäsie
der Technischen Universität
Mommsenstraße 13
01062 Dresden
Tel. 0351-463-4097
Fax: 0351-463-7019
http://astro.geo.tu-dresden.de

Institut für Theoretische Physik
der Johann Wolfgang Goethe-
Universität Frankfurt
(Astrophysik)
Max-von-Laue-Str. 1
60438 Frankfurt/Main
Tel. 069-798-47816
Fax: 069-798-47876
http://th.physik.uni-frankfurt.de/

Kiepenheuer-Institut für Sonnenphysik
Schöneckstraße 6
79104 Freiburg
Tel. 0761-3198-0
Fax: 0761-3198-111
http://www.kis.uni-freiburg.de

Max-Planck-Institut
für Astrophysik
Karl-Schwarzschild-Straße 1
85748 Garching
Tel. 089-30 000-0
Fax: 089-30 000-3235
http://www.mpa-garching.mpg.de

Max-Planck-Institut
für Extraterrestrische Physik
Giessenbachstraße
85 748 Garching
Tel. 089-30 000-0
Fax: 089-30 000-3569
http://www.mpe-garching.mpg.de

Technische Universität München
Lehrstuhl für Experimentalphysik und
Astro-Teilchenphysik
Department E15
James-Franck-Straße
85748 Garching
Tel. 089-289-12511
Fax: 089-289-12680
http://www.e15.physik.tu-muenchen.de

Member institutes of the German Council of Observatories

Georg-August-Universität Göttingen
Institut für Astrophysik Göttingen
Friedrich-Hund-Platz 1
37077 Göttingen
Tel. 0551-39-5042
Fax: 0551-39-5043
http://www.astro.physik.uni-goettingen.de/institute/index.de.html

Max-Planck-Institut
für Gravitationsphysik
Albert-Einstein-Institut
Am Mühlenberg 1
14476 Golm
Tel. 0331-567-70
Fax: 0331-567-7298
http://www.aei-potsdam.mpg.de

Hamburger Sternwarte
Gojenbergsweg 112
21029 Hamburg
Tel. 040-7252-4112
Fax: 040-7252-4198
http://www.hs.uni-hamburg.de

Max-Planck-Institut
für Gravitationsphysik
Albert-Einstein-Institut
Callinstr. 38
30167 Hannover
Tel. 0511-762-2229
Fax: 0511-762-5861
http://www.aei.mpg.de/english/metanavi/contact/address/index.html

Astronomisches Rechen-Institut (ARI)
Zentrum für Astronomie der Universität Heidelberg (ZAH)
Mönchhofstraße 12-14
69120 Heidelberg
Tel. 06221-54-1845
Fax: 06221-54-1888
http://www.ari.uni-heidelberg.de

Institut für Theoretische Astrophysik (ITA)
Zentrum für Astronomie der Universität Heidelberg (ZAH)
Albert-Ueberle-Str. 2
69120 Heidelberg
Tel. 06221-54-4837
Fax: 06221-54-4221
http://www.ita.uni-heidelberg.de

Landessternwarte Königstuhl (LSW)
Zentrum für Astronomie der Universität Heidelberg (ZAH)
Königstuhl 12
69117 Heidelberg
Tel. 06221-54-1700
Fax: 06221-54-1702
http://www.lsw.uni-heidelberg.de

Max-Planck-Institut
für Astronomie
Königstuhl 17
69117 Heidelberg
Tel. 06221-528-0
Fax: 06221-528-246
http://www.mpia.de/index.html

Max-Planck-Institut
für Kernphysik
Saupfercheckweg 1
69117 Heidelberg
Tel. 06221-516-295
Fax: 06221-516-549
http://www.mpi-hd.mpg.de

Astrophysikalisches Institut und
Universitäts-Sternwarte
der Universität Jena
Schillergäßchen 2
07745 Jena
Tel. 03641-947501
Fax: 03641-947502
http://www.astro.uni-jena.de

Max-Planck-Institut
für Sonnensystemforschung
Max-Planck-Str. 2
37191 Katlenburg-Lindau
Tel. 05556-979-0
Fax: 05556-979-240
http://www.mps.mpg.de/

Institut für Theoretische Physik u.
Astrophysik/Universität Kiel
Leibnizstr. 15
24 098 Kiel
Tel. 0431-880-4110
Fax: 0431-880-4100
http://www.astrophysik.uni-kiel.de

I. Physikalisches Institut
der Universität Köln
Zülpicher Straße 77
50 937 Köln
Tel. 0221-470-5737
Fax: 0221-470-5162
http://www.ph1.uni-koeln.de/index.html

Institut für Astronomie u. Astrophysik
der Universität München
Universitäts-Sternwarte
Observatorium Wendelstein
Scheinerstraße 1
81679 München
Tel. 089-2180-6001
Fax: 089-2180-6003
http://www.usm.uni-muenchen.de

Astrophysikalisches Institut Potsdam
An der Sternwarte 16
14482 Potsdam
Tel. 0331-7499-0
Fax: 0331-7499-200
http://www.aip.de

Astrophysik an der
Universität Potsdam
Postfach 601553
Am Neuen Palais 10
14415 Potsdam
Tel. 0331-977-1054
Fax: 0331-977-1107
http://www.astro.physik.uni-potsdam.de

4 pi Systeme Gesellschaft für
Astronomie und Informations-
verarbeitung mbH
Sternwarte Sonneberg
Sternwartestraße 32
96515 Sonneberg
Tel. 03675-812-10
Fax: 03675-812-19
http://www.stw.tu-ilmenau.de

Thüringer Landessternwarte Tautenburg
Karl-Schwarzschild-Observatorium
Sternwarte 5
07778 Tautenburg
Tel. 036427-863-0
Fax: 036427-863-29
http://www.tls-tautenburg.de

Institut für Astronomie u. Astrophysik
der Universität Tübingen
Abteilung Astronomie
Sand 1
72076 Tübingen
Tel. 07071-29-72486
Fax: 07071-29-3458
http://astro.uni-tuebingen.de

Institut für Astronomie u. Astrophysik
der Universität Tübingen
Abteilung theoretische Astrophysik
Computational Physics
Auf der Morgenstelle 10
72076 Tübingen
Tel. 07071-29-75468
Fax: 07071-29-5889
http://www.tat.physik.uni-tuebingen.de/

Lehrstuhl für Astronomie
Institut für Theoretische Physik und
Astrophysik
der Universität Würzburg
Am Hubland
97074 Würzburg
Tel. 0931-888 5877
Fax: 0931-888 4604
http://www.physik.uni-wuerzburg.de/einrichtungen/institut_fuer_theoretische_physik_und_astrophysik/

Table 6.1: Scientific fields of the astronomy and astrophysics institutes in Germany

Institute	Field(s)	Technology development
Bamberg, StW	White dwarfs	
	Close binaries	
	Hot stars	
Berlin, DLR	Star formation and interstellar medium	Yes
	Planetary reconnaissance	
Berlin, TU	Stellar atmospheres	
	Circumstellar dust envelopes	
	Astrochemistry	
	Numerical astrophysics	
Bochum, AIB	Dwarf galaxies	Yes
	Interstellar medium	
	Star formation	
Bochum, Theor. physics	Plasma physics	
	High energy physics and space research	
Bonn, IAEF	Gravitational lenses	
	Cosmology	
	Interplanetary space	
	Upper atmosphere	
Bonn, MPIfR	Star formation	Yes
	Late stages of stellar genesis	
	Active galactic nuclei	
	Galaxies and their evolution	
	Magnetic fields	
	Pulsars	
	Astrochemistry	
	Cosmology	
Bonn, RAIUB	Dwarf galaxies	
	Galaxy evolution	
	Interstellar medium	
	Submillimetre and millimetre astronomy	
Bonn, StW	Stars, stellar populations	Yes
	Structure of the galaxy	
	Interstellar UV spectroscopy	
	Dwarf galaxies	
Dresden, Univ.	Solar system	
Frankfurt, UF	Interstellar matter	
	Radiation transport, molecular lines	
	Quasars	
Freiburg, KIS	Solar physics	Yes
	Stellar activity	
	Magnetohydrodynamics	

Member institutes of the Council of German Observatories

Table 6.1: continued

Institute	Field(s)	Technology development
Garching, MPA	Nuclear astrophysics High-energy astrophysics Numerical astrophysics Stellar evolution Cosmology Structure and evolution of galaxies	Yes
Garching, MPE	Cosmology Final stages of stellar evolutions Black holes, active galactic nuclei Galaxy evolution Star and planet formation, extrasolar planets Astrophysical plasmas	Yes
Garching, TUM	Astroparticle physics	
Göttingen, USG	Theoretical stellar dynamics Numerical astrophysics High-energy astrophysics Stellar astronomy Galaxies and galaxy clusters The sun (observation and theory)	Yes
Hamburg, UH	Cosmology Stellar activity and coronae Final stage of stellar evolution Stellar atmospheres	Yes
Hannover, Univ.	Gravitational waves	Yes
Heidelberg, ZAH (ARI)	Astrometry, stellar dynamics Ephemerides and calendars Bibliography	Yes
Heidelberg, ZAH (ITA)	Stellar atmospheres Accretion disks Genesis of stars and planets Numerical astrophysics	
Heidelberg, ZAH (LSW)	Active galaxies and quasars X-ray sources Novae Stellar genesis Jets and disks Black holes Hot stars The sun	Yes

Table 6.1: continued

Institute	Field(s)	Technology development
Heidelberg, MPIA	Structure and evolution of galaxies Star and planet formation, Interstellar medium Dark matter Black holes Numerical astrophysics	Yes
Heidelberg, MPIK	TeV gamma astronomy Astroparticle physics Plasma astrophysics Infrared astrophysics	Yes
Jena, AIU	Star and planet formation Interstellar medium Laboratory astrophysics, astrochemistry Numerical astrophysics	Yes
Katlenburg-Lindau MPAE	Solar physics Heliosphere physics Space plasmas and magnetohydrodynamics Sun-Earth relationships	Yes
Kiel, Uni	Stellar atmospheres White dwarfs Cool stars and the sun Interstellar medium Stellar dynamics Galaxy evolution	
Cologne, 1st Phys. Inst.	Star formation Galaxy evolution Active galaxies Interstellar medium	Yes
Munich, USM	Stellar winds and atmospheres Structure and genesis of galaxies Plasma astrophysics Binaries Cataclysmic variables	Yes
Potsdam, AIP	Magnetohydrodynamics Physics and activity of stars Solar physics Star formation and extrasolar planets Formation and evolution of galaxies Cosmology Numerical astrophysics	Yes

Member institutes of the Council of German Observatories

Table 6.1: continued

Institute	Field(s)	Technology development
Potsdam, UP	Hot stars	Yes
	Gravitational lenses	
	Extrasolar planets	
	Quasars	
Potsdam, AEI	Gravitational waves	
	Quantum gravity	
	Numerical gravity	
Sonneberg, StW	Variable stars	Yes
	Near-Earth asteroids	
	Sky surveys	
	Databases	
Tautenburg, LSW	Extrasolar planets and brown dwarfs	
	Star formation	
	Active galaxies	
	Gamma ray bursts	
	Stellar pulsation	
Tübingen, IAAT	Final stages of stellar development	Yes
	Experimental and theoretical X-ray astronomy	
	UV astronomy	
	Relativistic astrophysics	
	Star and planet formation	
	Numerical astrophysics	
Würzburg, Univ.	Extragalactic astronomy	
	Astroparticle physics	
	High-energy astrophysics	

Table 6.2: Scientific personnel at German institutes

Personnel	Total institutes	University	Non-university institutes (incl. MPG)
Band C4 professors	50	27	23
Band C3 professors	47	19	28
Established posts Band C2 and C1, Band A/Federal Employees' Collective Salary Agreement	375	94.5	280.5
Third party funding	202	83	119
Total	674	223.5	450.5
Doctoral candidate			
Established posts	147	31	116
Third party funding	262	215	47
Total	409	246	163
Graduate students	215	163	52

Table 6.3: Development of positions in astronomy with time

Year	Established posts	Third party funding	Technicians	Administration
1962	100	30	130	17
1987	379	109	342	112
1999	375	202	405	90.9

Table 6.4: Development of completed doctoral theses and diploma theses

Year	Graduate theses	Percentage	Doctorates	East/West	Percentage
1991	253	7.5%	55	1/54	4.8%
1993	207	5.8%	61	3/58	4.3%
1995	318	10.6%	50	1/24	3.3%
1997	304	8.9%	83	6/77	5.5%
1998	283	8.8%	88	9/79	5.9%
1999	215	8.4%	77	8/69	5.0%

The percentage proportion is in terms of the total of all diploma theses and dissertations in the physics faculty for the respective year.

Table 6.5: Development of the number of publications and conference proceeding with time

Year	Refereed journals	Conference proceedings
1991	1435	244
1993	1646	379
1995	1277	886
1997	1564	894
1998	1726	1034
1999	1681	1108

Source: Mitteilungen der Astronomischen Gesellschaft (Transactions of the Astronomical Society)

Statistical material on the development of astronomy and astrophysics

The following institutes are included in the statistics:

Bamberg, Berlin, Bochum (TH, AI), Bonn (IAEF, RAIUB, StW, MPIfR), Dresden, Frankfurt, Freiburg (KIS), Göttingen, Hamburg, Hanover, Heidelberg (ARI, ITA, LSW, MPIA, MPIK), Jena, Kiel, Cologne, Katlenburg-Lindau, Munich, Garching (TUM, MPA, MPE), Potsdam (AIP, Univ.), Sonneberg, Tautenburg, Tübingen, Würzburg.

Fig. 6.1: Distribution of established scientific posts in astronomy in Germany

Glossary/Acronyms

Organisations/Institutes

AI	Astronomisches Institut, Bochum (Astronomical Institute, Bochum)
AIP	Astrophysikalisches Institut Potsdam (Astrophysical Institute Potsdam)
AIU	Astrophysikalisches Institut und Universitätssternwarte, Jena (Astrophysical Institute and University Observatory, Jena)
AWG	ESA Astronomy Working Group
ARI	Astronomisches Recheninstitut, Heidelberg (Astronomical Computing Institute, Heidelberg)
BMBF	Bundesministerium für Bildung und Forschung (Federal Ministry of Education and Research)
CARA	California Association for Research in Astronomy
CNRS	Centre National de la Recherche Scintifique
DFG	Deutsche Forschungsgemeinschaft (German Research Foundation)
DLR	Deutsches Zentrum für Luft- und Raumfahrt (German Aerospace Centre)
DSAZ	Deutsch-Spanisches Astronomisches Zentrum (German-Spanish Astronomical Centre)
ESA	European Space Agency
ESO	European Southern Observatory
EU	European Union
FrInGe	Frontiers of Interferometry in Germany (Deutsches Interferometriezentrum für den optischen und infraroten Wellenlängenbereich, Heidelberg)
IAEF	Institut für Astrophysik und Extraterrestrische Forschung der Universität Bonn (Bonn University Institute for Astrophysics and Extraterrestrial Research)
IAU	International Astronomical Union
IGN	Instituto Geografico Nacional, Spain
INSU	Institut National des Sciences de l'Univers, France
IRAM	Institut de Radioastronomie Millimetrique, Grenoble, France; Granada, Spain
LNGS	Laboratorio Nazionale del Gran Sasso
LSW	Landessternwarte, Heidelberg (State Observatory, Heidelberg)
MPE	MPI für Extraterrestrische Physik, Garching (MPI for Extraterrestrial Physics, Garching)
MPA	MPI für Astrophysik, Garching (MPI for Astrophysics, Garching)
MPI	Max-Planck-Institut (Max Planck Institute)
MPIA	MPI für Astronomie, Heidelberg (MPI for Astronomy, Heidelberg)
MPIfR	MPI für Radioastronomie, Bonn (MPI for Radio Astronomy, Bonn)
MPIK	MPI für Kernphysik, Heidelberg (MPI for Nuclear Physics, Heidelberg)

MPG	Max-Planck-Gesellschaft (Max Planck Society)
NAS/NRC	National Academy of Science
NASA	National Aeronautic and Space Agency
PPARC	Particle Physics and Astronomy Research Council, UK
RAIUB	Radioastronomisches Institut der Universität Bonn (Radio Astronomy Institute of the University of Bonn)
RDS	Rat Deutscher Sternwarten (Council of German Observatories)
SFB	Sonderforschungsbereich (Collaborative Research Centre)
Stw	Sternwarte (Observatory)
SSAC	ESA Space Science Advisory Committee
Uni	Universität (University)
USM	Universitätssternwarte München (Munich University Observatory)
VBF	Verbundforschung (Cooperative Research)
ZAH	Zentrum für Astronomie der Universität Heidelberg

Astronomical and other terms

AGN	Aktiver Galaxienkern (Active Galactic Nucleus)
EDV	Elektronische Datenverarbeitung (Electronic Data Processing, EDP)
ESFON	European Star Formation Network
EUV	Extreme Ultraviolet
GAVO	German Virtual Observatory
GCT	Gregory-Coude-Teleskop
GRID	Nachfolge des Internet (Successor to the Internet)
GRK	Graduiertenkolleg (Research Training Group)
IC	Index Catalogue
IR	Infrarot (Infrared)
ISS	International Space Station
KIS	Kiepenheuer-Institut für Sonnenphysik (Kiepenheuer Institute for Solar Physics)
MACHOs	Massive Compact Halo Objects
MCG	Morphological Catalogue of Galaxies
MeV	Mega-Elektronenvolt (Mega Electron Volt)
Mrk	Markarian
MIDEX	Medium-Class Explorer
NGC	New General Catalogue
NIR	Near Infrared
NORAS	Northern Rosat All-Sky (Cluster Survey)
OPTICON	Optical Infrared Coordination Network for Astronomy
QSO	Quasistellar Object
REFLEX	ROSAT/ESO Flux Limited X-Ray Survey
SDSS	Sloan Digital Sky Survey
SMEX	Small Explorer
SPP	Schwerpunktprogramm (Priority Programme)
SZ-Effekt	Sunyaev-Zeldovich-Effekt (Sunyaev-Zel'dovich Effect)

TeV	Tera-Elektronenvolt (Tera Electron Volt)
THz	Tera-Hertz
TMR	Training and Mobility of Researchers
ULIRGs	Ultra Luminous Infrared Galaxies
UV	Ultraviolett (Ultraviolet)
VHE	Very High Energy
WIMPs	Weakly Interacting Massive Particles

Telescopes, instruments, experiments

Gamma range

CGRO	Compton Gamma Ray Observatory
COMPTEL	Imaging Compton Telescope
EGRET	Energetic Gamma Ray Experiment Telescope
INTEGRAL	International Gamma Ray Laboratory
GLAST	Gamma Ray Large Area Space Telescope
HEGRA	High Energy Gamma Ray Astronomy
H.E.S.S	High Energy Stereoscopic System
MAGIC	Major Atmospheric Gamma Imaging Cherenkov Telescope
MEGA	Medium Energy Gamma Ray Astronomy

X-ray range

ABRIXAS	A Broad Band Imaging X-Ray All-Sky Survey
ASCA	Advanced Satellite for Cosmology and Astrophysics
Chandra	American X-ray satellite (formerly AXAF), named after S. Chandrasekhar
EPIC	European Photon Imaging Camera
ROSAT	X-ray satellite
ROSITA	X-ray Survey with Imaging Telescope Array, successor to ABRIXAS
eROSITA	extended X-ray Survey with Imaging Telescope Array, Telescope on Russian SRG-Mission
XEUS	X-Ray Evolving Universe Spectroscopy Mission
XMM-Newton	X-Ray Multi Mirror Mission, ESA X-ray satellite, named after Isaac Newton

UV, optical and infrared range

ALFA	Adaptive Optics with a Laser For Astronomy, Calar Alto
ASTRO-SPAS	Shuttle Pallet Satellite, reusable space platform
ATST	Advanced Technology Solar Telescope
BUSCA	Simultaneous four-colour CCD camera, Calar Alto
COS	Cosmic Origins Spectrograph
Darwin	Infrared Space Interferometer, named after C. Darwin
DENIS	Deep Near Infrared Survey
Eddington	Asteroseismology mission, named after A.S. Eddington
FIRST	Far Infrared Space Telescope, now Herschel
Herschel	New name for FIRST, Far Infrared Space Telescope
FORS 1 + 2	Focal Reducer and Spectrograph (two instruments)
FUSE	Far Ultraviolet Spectroscopic Explorer

GAIA	Global Astrometric Interferometer for Astrophysics
GREGOR	Solar telescope, Tenerife, named after J. Gregory
HET	Hobby Eberly Telescope, Texas
HHT	Heinrich Hertz Telescope
HIPPARCOS	Astrometry satellite, named after Hipparchos
HST	Hubble Space Telescope
IRAS	Infrared Astronomical Satellite
ISO	Infrared Space Observatory
ISOCAM	ISO Camera
ISOPHOT	ISO Imaging Photopolarimeter
IUE	International Ultraviolet Explorer
JWST	James Webb Space Telescope, formerly Next Generation Space Telescope, NGST
KAO	Kuiper Airborne Observatory
Keck-Telescope	Two 10 m telescopes on Mauna Kea, Hawaii, named after Keck
LASCO	Large Angle and Spectrometric Coronograph
LBT	Large Binocular Telescope, Mt. Graham
LBTI	LBT interferometer
LUCIFER	LBT NIR Spectrograph and Integral-Field Unit
2MASS	2 Micron All Sky Survey
MIRI	Mid Infrared Instrument
MONET	Monitoring Network of Telescopes
NGST	see JWST
NTT	New Technology Telescope, La Silla
OmegaCAM	Camera for the VST
ORFEUS	Orbiting and Retrievable Far and Extreme Ultraviolet Spectrograph
OWL	Overwhelmingly Large Telescope
PRIME	Primordial Explorer
SALT	Southern Africa Large Telescope
SHARP	High-resolution infrared camera
SIRTF	Space Infrared Telescope Facility
SOFIA	Stratospheric Observatory for Infrared Astronomy
SOHO	Solar and Heliospheric Observatory
Solar Orbiter	Solar research space probe
STEREO	Stereoscopic View of the Sun-Earth Connection
STELLA	Robotic telescope for investigating stellar activity, Tenerife
SUNRISE	Solar research balloon experiment
THEMIS	Solar telescope, Tenerife
TIMMI-2	Thermal Infrared Multimode Instrument on the 3.6 m telescope, La Silla
ULYSSES	NASA solar pole mission
VISTA	Visible and Infrared Survey Telescope for Astronomy
VLT	Very Large Telescope, Paranal
VLTI	VLT Interferometer
VST	VLT Survey Telescope, Paranal

VTT	Vakuum-Turmteleskop, Teneriffa (vacuum tower telescope, Tenerife)
WSO	World Space Observatory

Submillimetre radio range

ALMA	Atacama Large Millimetre Array, Chajnantor
APEX	Atacama Pathfinder Experiment; Pilot-Experiment for ALMA
BOOMERanG	Balloon Observations of Millimetric Extragalactic Radiation and Geophysics
CARMA	California Millimeter Array
COBE	Cosmic Background Explorer
EVN	European VLBI Network
EVLA	Expanded Very Large Array
JCMT	James Clerk Maxwell Teleskop, Mauna Kea (telescope)
LOFAR	Low Frequency Array, planned radio telescope
MAMBO	Max-Planck-Millimeter-Bolometer (Max Planck Millimetre Bolometer)
MAXIMA	Millimeter Anisotropy Experiment Imaging Array
MERLIN	Multi-Element Radio Linked Interferometer Network
MMIC	Monolithic microwave integrated circuits
Planck	Cosmic Background Mission, named after Planck
SCUBA	Submillimetre Common-User Bolometer Array
SKA	Square Kilometre Array, planned radio telescope
VLA	Very Large Array
VLBA	Very Long Baseline Array
VLBI	Very Long Baseline Interferometer
WMAP	Wilkinson Microwave Anisotropy Probe

Astroparticle physics

AMANDA	Antarctic Muon and Neutrino Detector Array
ANTARES	Astronomy with a Neutrino Telescope and Abyss Environmental Research
AUGER	Cosmic Ray Project, named after P. Auger
BOREXINO	Detector for Low Energy Solar Neutrinos, Gran-Sasso-Laboratorium (Gran Sasso Laboratory)
CDMS	Cryogenic Dark Matter Search
CRESST	Cryogenic Rare Event Search with Superconducting Thermometers
DAMA	Dark Matter searches with low activity scintillators, Gran-Sasso-Laboratorium (Gran Sasso Laboratory)
EDELWEISS	Experience pour Detecter les WIMPs en Site Souterrain
EUSO	Extreme Universe Space Observatory

GALLEX	Gallium Experiment, Gran-Sasso-Laboratorium (Gran Sasso Laboratory)
GENIUS	Germanium in liquid Nitrogen Underground Setup
GNO	Gallium Neutrino Observatory, Gran-Sasso-Laboratorium (Gran Sasso Laboratory)
Homestake	Neutrino experiment in the American Homestake Mine
ICECUBE	Neutrino-Teleskop: Instrumenting a cubic kilometre of ice under the South Pole (telescope)
KASCADE	Karlsruhe Shower Core and Array Detector
LENS	Low Energy Neutrino Spectroscopy
SAGE	Sovjet-American Gallium Experiment
SNO	Sudbury Neutrino Observatory
Super-Kamiokande	Japanese neutrino detector

Gravitation

GEO600	Laser interferometer in Hannover with 600 m arm length
GP/B-Mission	Gravity Probe B
LIGO II	Laser Interferometer Gravitational-Wave Observatory
LISA	Laser Interferometer Space Antenna
STEP	Satellite Test of Equivalent Principle
VIRGO II	Italian-French laser interferometer